OVERCOMING ... SUR
OBSTACLES (crossword grid spelling out:)

OVERCOMING · SURVIVE · PAIN · CEREBRAL · VICTORY · REACHEEBUILD · DISEASE · KIDNEY · GLUCOMA · GOAL · MOBILITY · DYSLIS... · TRAVEL · HOPE · BELIEF IN YOURSELF · FAILFFIFF · TRANSPLANT · LOVE · REALITY · TEALITY

A LIFETIME OF SUCCESS

BY JAMES A SCHANER

Copyright 2012

ISBN: 978-1-105-88696-6

Acknowledgments

This book is dedicated to the many professionals, medical staff, dialysis staff and family members who have gained my trust in a lifelong battle to overcome several challenges. Thank you to the teachers, professors and those who initially did not have an understanding of just how hard a man would fight. In time they too have come to know just what it takes to conquer challenges. I hope that somehow I have inspired others to do the same. It's not easy but it can be accomplished if you only believe in your own potential.

Thanks to my former employers who let me volunteer in order to pick up the necessary skills to do a job for pay. Without them, the career path I took would not have been possible. It is my hope that by sharing these experiences employers will be able to learn from the challenges I had to face each day.

Special thanks to my father for the selfless gift of a kidney. Without it I would not be alive today! My mother deserves a great deal of credit for teaching me the tools necessary to fight juvenile diabetes and ultimately defeat it.

Thanks to Deb Harvey and all the employees who aided me in turning a pipe dream into reality. Finally I want to thank Chuck, Patty, Mark and all of you with Independent Life Services in Queen Creek, Arizona. You allowed a man with disabilities and challenges of his own to open people's eyes wide enough to learn that all one needs is a moment in time where they are more than they thought they could be.

Table of Challenges

The Early Years: 1967-1973

On February 23, 1967, I made my initial appearance into the world. This would have been normal under any circumstance but this one. It seems that mother-nature was ready for an early appearance. On this Saturday evening the writer arrived weighing 8 pounds, 10 ounces, and approximately 6 weeks ahead of schedule. This happened because my mother had been fighting Type I Diabetes for approximately thirteen years and simply could not carry me to term. The results of this early birth could have been catastrophic. I stopped breathing three-times following delivery. In the end my parents knew that I had Cerebral Palsy. They did not know the damage to come from it. Quoting my mother, "The doctor told me that in light of the traumatic pregnancy I should not have any more children." My mother was not a quitter. Two years later my parents adopted the first of what would ultimately be five additional siblings into my family. I should have known right then where my attitude came from. If you want to motivate me all you need to do is tell me no.

Over the next few years my parents were constantly on the watch. Given what they were told who could blame them. They were to look for possible developmental delays which would give them a minor look at the price tag paid by me for nature's mistake. For me much of this would be discovered viewing home movies. Among the things I saw were attempts to walk while holding onto walls and furniture for balance. That first step on my own did not take place until eighteen months of age when my father held a

Coke bottle within sight as a prize. After a few more weeks I could finally walk on my own.

Time moved on with the ever-present reminder for my parents that what was to come was unknown. One summer morning in 1971 the truth was released. I often visited physicians at the University of Michigan Medical Center in Ann Arbor. Over time these specialists evaluated me and eventually came up with a less than thrilling forecast for my future. Had I been old enough to understand tears would have flowed freely as they have at times over the years. Based on testing by the experts at Ann Arbor they told my parents I would probably not advance beyond a sixth grade education level. That any hopes for future employment were minimal. And lastly, but not least, they told Mom and Dad that I should never marry. I must say that proving them wrong on two out of three isn't bad. More on these two as we progress.

In hindsight I must realize that the advice given was based on 1971 thinking. Today we're much better off in terms of people's ability to look beyond what they see. If these professionals had seen what I was doing today shock would set in. Attitudes and our society have also changed. While there are still pockets of the old thought process today, in general, opportunities for disabled Americans are far greater than they were back in 1971. I had to earn my breaks along the way. Although at first I did not feel comfortable with it I am glad I was given a platform to pick up a skill before receiving the responsibility of a paid job.

During those early years there were many attempts to maintain my physical strength. These were not appreciated much by a four-year-old boy, and continued well into his childhood years. Many a night as a youth I spent the evening in braces that came up to shin level just below the knee. The pain of wearing these braces during bedtime hours as a growing four-year-old boy left me crying on many occasions as well as kicking holes in walls either to the side of my single twin bed or in the wall between the lower and upper bunks. I felt like exercise was being forced upon me when I should have been thinking it was for my own good. My parents tired of pressuring me over the years and relented. I know that was my mistake rather than theirs. For a short while in junior high school we tried another form of brace. Twister cable braces that were connected to regular shoes at the bottom and metal cables wrapped in plastic which ran up the side of the leg to the hip. By then the ability to walk had been established and my body's own strength caused me to "walk right through" the therapeutic effects of the device.

In the autumn of that year I began my education. Unlike the majority of children my age my first years were spent in special education. Although I did not know at the time, this was a school that specializes in education of disabled children. For a starting point this was actually a very nice move. It allowed me to socialize with people on my level. All of my peers were disabled in one form or another. Most were in wheelchairs, on crutches, walking with limps or carrying other deformities. As I grew older I became more and more self-conscious of my own deformity. The disability I have focuses a little bit on each part of my body. Predominantly, it has left my feet pigeon-toed. (The feet turn in approximately 30°) It also led me to walking on my toes while most people do so heel-to-toe. This really stood out in physical exercise classes. I was not able to do push-ups or chin-ups like my peers. I could not play sports that involve movement of the feet and contact of objects with the same. For me there was not only physical education but occupational therapy. These exercises focused on coordination with hand and eye.

Despite all of these challenges however I proceeded to make friends. I also learned some of the basics of human communication. I can remember conversations with my kindergarten teacher. It was here that I learned the skills of saying please and thank you. I had already learned from my parents the alphabet, numbers from 1 to 10 and later 11 to 20. Simple addition and subtraction came next. This trend continued right through the end of the first grade. As I continued to grow the speech that was so important in any man's life came along with it.

By the spring of 1974 my parents had begun to see the success I discussed earlier. They conferred with school staff about mainstreaming me. In layman's terms this means being placed with other students in normal education circumstances. They ran into an attitude which told them if they do so they would be leaving their child emotionally shortchanged because he would not grow as fast as his peers. They felt that this would lead to further isolation over time. My parents' answer was to have me tested according to state standards set with every other student of my age. I don't remember the advice I was given by my parents but I do know that I did very well on that test. The outcome was enough for the school staff and the district to change its mind in the matter. Beginning in the Fall of 1974 I would be in school with other children my age. This is a pattern that stayed true and remains so to this day.

My Family and I at Rocky Point Mexico

Victories and Setbacks Aplenty: 1974-1987

In the Fall of 1974 I was ready for my education to start again. Unlike the previous school this one was just blocks from my home. Although I could not wait for this school year to begin I was eager to enjoy a well-deserved summer vacation. Often neighborhood children were visited. Activities included playing dice games, board games, and card games. Play generally included horseplay with my siblings. It was all in good fun and it generally toughened me for later years. A favorite thing to do was to meet at the corner of our subdivision where there was an open field. In this open field softball games would be a regular occurrence in the summertime.

One of my closest neighborhood friends was a boy who discovered I had a real interest in radio. We also shared an interest in professional sports. At this young age, I quickly learned the art of imagination. Listening to play-by-play men broadcast Detroit Tigers baseball games quickly became a hobby of mine. These men could easily have filled a gap left by a busy father or mother who could not tell their child bedtime stories. This is not to say that my parents would not read me bedtime stories, but when they couldn't, these men were my heroes.

If the season was winter, we could ice skate on a frozen ditch. There were many days when I remember sliding on the ice in winter boots. I tended goal in front of the pipes which bordered the walls of that ditch when I played pond hockey. Our home also had a basement, which allowed me to play indoor sports such as table tennis, darts and billiards. I was also able to bring the hockey experience downstairs with a tabletop version of ice hockey. These games really improved my hand eye coordination. Because of my height and size advantage, touch football could be played with relative ease.

There was an awful lot of time in those years, but in this particular summer, one more heavy hand of fate was placed upon me. This happened innocently enough when my family and I decided to get involved in foreign adoption through a potluck held in the town of Lapeer, Michigan. It was here that we continued to learn how to bring my middle brother to the United States. It was a very hot, humid day. For some reason unknown to me, I was continually thirsty. Seemingly, there was no amount of fluid that could satiate a need for water. I thought it was just a strange day and it would pass in time. As we began our journey home, a distance of some 100 miles plus, I began to need to expel the spoils of that meal. This would be normal in most circumstances until you realize that you are doing it once every 15 minutes! After several stops along roadsides, my mother finally told my father: "Dick, why don't you take Jim to the bathroom. I'll call the hospital. I already know what he's got."

A quick trip to the hospital confirmed my mother's concerns. On their advice, I was taken to Ann Arbor. I spent the next week at Mott Children's Hospital. I was now a Type I Diabetic. For the next seven days I would have to adjust to a new diet, new exercise program after hospitalization, and a whole other meaning of the term "different". I also quickly learned a different destination of pain. In those days, blood sugar results would not come in 10 to 30 seconds. It often took up to three hours for these results to come back. There were no instant accuracy blood sugar checks then. Instead, a patient received 14-gauge needles in the vein between my forearm and elbow. It meant insulin shots at meals. These usually came prior to breakfast. At that time, medical social workers were relatively reserved to an adult patient.

This period in my life affected me greatly on an emotional level. Dietitians would bring me menus asking me to pick my own choices. When the menus came back with check marks and red lines, replaced by the healthier choices needed for a diabetic I could not understand why. At one point I was brought a snack which was required as part of the new diet but did not include things that a seven-year-old boy would necessarily choose. Instead I would receive diet ginger ale, cheese and crackers. I became so angry once that I shook the content of the soda can and sprayed it on the ceiling of my room. I realized that I was fighting a lost cause. My only alternative was to turn to my parents who repeatedly came to me through this adjustment. I can remember to this day my parents telling me that if I did not follow the advice given I would spend more time in the hospital, become sicker, and not feel well. I thank my parents for that advice because without it I would never have made it as far as I have.

The treatment strategy for diabetes also has changed over time. When I was first diagnosed the strategy was for a child of my age to partake in a couple of snacks between meals. One snack occurred between lunch and dinner and the other between dinner and breakfast the next morning. While each diabetic case differs most young children are still growing. The last thing my dietitian wanted to do was stunt that growth before it was completed. Once adulthood arrives the disease itself generally worsens. At that time it was no longer necessary to eat snacks between meals.

All the news is not necessarily bad. They did allow me to break the diet on special occasions such as holidays, birthdays, and some family gatherings. These were called "hog wild days." You might say that this was the beginning of the current diabetic diet which is not quite as strict with respect to the "tight ship" theory of the past. The change was made possible by advancements in medical technology. Among the improvements is the fact that you can now determine your blood glucose levels with almost clinical accuracy as fast as five seconds. There are also four different types of insulin on the market today. In addition to those, for some diabetics of age, there are now pills which are used to control blood sugar level. There are even monitors to control A1C levels, a previously unknown, but important value which will help patients determine their risk for cardiac illnesses. These are complications for any diabetic. If it's medical science itself that you wish to discuss I can't help you there. It is now possible for a diabetic of a certain age to wait for not only a kidney transplant but a pancreas as well! This advancement in transplants was nothing more than a pipe dream for most of my battle. Critics would rather I not talk about transplants when I go to diabetic charity walks. It seems that they would rather have a mass cure for everyone. There is nothing wrong with that stance. What I tell people is the choice is entirely up to them. If they don't want to do what it takes to get this life-saving surgery it is their choice. There is more on this to come as our book progresses.

Shortly after leaving the hospital I began to prepare for life as a second grader. There were some changes that had to be made due to this disease but they were not of such a serious nature that I couldn't manage. Among these changes were that as a youngster I carried sugarless candy with me. This was to be stored in the teacher's desk. It would be used in the place of birthday cakes and other such things that students would have on their birthdays. A second and even more important tool was a 6-pack of miniature sized canned juices. These were used to prevent low blood sugar. In the event that things got really low there was also the dreaded standby of Glucagon. In any case when the sugar dropped low it became my responsibility to react to it. If I was low a can of orange juice would usually do the trick. In the case of a severe low I would be sent to a school nurse who had the same juice. I brought it and left it with her or she had juice herself in the office. Glucagon never had to be used within the school setting in my life. When you learn lessons from your own mother regarding treatment of such episodes it becomes clear that you have an ally at home not just a mom.

What made the diabetic challenge more difficult was the fact that I also had Cerebral Palsy. I often refer to this double whammy as genetic dynamite. It is hard enough to carry either disease by itself but when you carry both of them together it is very important that you keep your brains about you. Many times things happened where one would mix with the other to the point where I did not know I was having a reaction to insulin. In these instances it was imperative that the people who cared for me during the day knew exactly what to do. For a few years I actually wore a medical alert chain around my neck for just such an emergency.

As the visits to Ann Arbor continued the primary focus of medical care shifted from Cerebral Palsy to Diabetes. It was not the fact that Cerebral Palsy is not necessarily covered but simply that Cerebral Palsy by itself could not kill me. The other disease very well could in its own right. Because of this combination doctors often allowed me to run a slightly higher blood sugar level. All told they went slightly outside the box because they would rather see me healthy then suffer additional brain damage. Now, when they were examining my lower extremities, instead of just looking at the deformity itself they were looking for signs of diabetic damage. It is for this reason as much as any other that I have continued to have my lower extremities evaluated. This has gone on now for the better part of forty years. When I asked for the foot check in dialysis clinics staff looked at me like I was crazy. But it is what they don't know that causes this. I do not ask this of them to be satisfied. This was now and always will be listed under self-preservation in my book. Better to find a problem out early than to take it home and assume that everything is okay. What they catch may be something more important than what my father could catch or what I can catch in my home. One of the least-anticipated items to look forward to is an imminent return to the hospital with a transplant on the line only to be sent home because I had a foot infection or sore that could have been healed if it were found soon enough. This whole controversy was caused by a mistake in medical records at the clinic from the start. Their records had me listed as a diabetic. The funny thing was the disease was cured before I ever showed up. Yet they did foot checks on me for four years before they realized their mistake. Now that it is necessary they don't perform them on non-diabetic patients. Policy is policy but now the need is even more apparent than it was then. This issue has been resolved with the passage of time. I have now retained my own podiatrist in private practice.

Back to the education front is where we proceed. Not only were my recesses quite different but lunch hour, physical education classes, and general socialization skills improved as well. When one is finally free to communicate with the regular kids the boost to his self-esteem is astounding! By the end of the fourth grade I had advanced so much that there was almost nothing I could not handle without help. Over this time I did take my academic lumps on at least one occasion when they began to grade penmanship as a separate category. Also, I was never a very good science student. I now wish that I had placed more focus on that particular subject as blood chemistry has now become such an important part of my day-to-day life.

I'm going to move quickly now to the summer of 1977. At this point the Michigan economy was in shambles. The result of high unemployment, higher taxes than here in Arizona and a lackluster education fund lead our state to severe cutbacks within education itself. We now see these things happening in Arizona schools quite a bit much for the same reason. When you combine these two factors with the fact that my mother had been a diabetic for more than two decades, that my mother had a son who is also diabetic and a cold climate that comes with Michigan in the wintertime, it became clear to my parents that it was time for us to move to greener pastures. Knowing this my father sought employment out-of-state. He had married an Arizona woman. She came to Michigan years earlier because of family circumstances beyond her control. Often the family vacationed in Arizona and loved it very much. Eventually they came to a decision that it may be best for all of us to move to Arizona. We put our family home on the market and it sold rapidly. When the house sold it felt like a part of our lives was changing forever. All was not lost because of a special move made in consultation with my parents and a fourth-grade teacher who cared enough to try and ease any perceived feeling of abandonment. This teaching hero of mine had the foresight to invite me to watch during the summer of that year. We discussed the situation, and she left me with the following advice: "Jim, you know as well as I do that you made a lot of friends here. I don't want you to forget about them but you must realize that as you leave your friends here you will make new ones and a whole lot more of them in Arizona." It was this advice that made the move as easy as possible for a growing boy.

On August 16, 1977, we left Michigan as residents. One of the only reasons I remember this date is because sometime that day Elvis Presley had passed away and it was brought up at the breakfast table. For the next three to four days we traveled packed in a Ryder moving truck and our car. We were headed for a new home somewhere in the metropolitan Phoenix area. The day we hit the outskirts of town I can remember the "Singing Weatherman" on local radio talking about how it was *only 114° outside!* I remember pleading with my father: "*why?*" His reply could not have been better spoken. "Jim, ask me this again in about four months." December came and I did not have to ask my father. It may have been 25° outside that morning when I left for school but I do not remember a single winter morning where I did not have to deal with snow as I had to in Michigan! Another advantage to this move was the fact that the elementary and junior high schools were adjoining buildings and the schoolyard in which I played while at school was directly across the street through a chain-link fence. This was paradise for a boy like me.

Within one year of that move there was a swimming pool in our backyard complete with a diving board and a large enough backyard that I could play tetherball, whiffle ball, or any other type of ball I wished. Within this yard were seven citrus trees. These were trees that when I was old enough I eventually earned the responsibility of watering. The front yard contained a driveway with a basketball goal on the roof.

That first year in school was quite different from any I had experienced in the past. Because of a massive period of growth within our district we were forced to attend double-sessions. The first of these ran from 7:30 AM to 12:30 PM, with the second running from 1 PM to 5:30 PM. This meant that the afternoon sessions had less recess time than the morning. Though this was true I could care less. I had the advantage of sleeping in late in the morning provided homework was done the night before. Being well-rested outweighed extra recess. In our school district money had to be readily available because people like us were flocking into the Phoenix area to begin a new phase of their lives. As a result music, art, and extracurricular activities were not threatened-as they were in Michigan.

Being in an entirely new climate did not seem to affect my academic performance. I was still an average to above average student. What I had to watch for was diabetic symptoms. Extra heat during the warm weather months left me vulnerable to decreased blood sugar levels and eating incorrectly may lead to hyperglycemic episodes.

The following year there was another move. This one involved a walk of about another block in length up the sidewalk to Desert Shadows Middle School. I would spend the next three years of my academic career in this building. Unlike elementary school junior high began the transition to high school-in the sense that every individual subject was taught in a different classroom. Even my homeroom class in the early morning was a separate room. This teacher was responsible primarily for taking role, allowing time for the administration to deliver announcements to students and staff, and preparing students for the day ahead. We would take the same subjects that we had before. The only difference would be that each course was taught by a different instructor in different classrooms. One advantage I had was that each year my homeroom teacher was a teacher of one of my subjects. This may have been an advantage to everyone I just did not know any better at the time. When you are dealing with different classes on an individual level you may not have the same classmates in one as you might another.

Each class lasted roughly 45-minutes with a 6-minute break between classes. This would allow a student to change classes from one room to another. My favorite new courses came in the latter two years of this experience. In seventh and eighth grade respectively I enrolled in Spanish language class. I also had an opportunity to enroll in industrial trade classes like wood and metal shop. One of my favorite courses was physical education. This hour of exercise provided a perfect release from the academic tension of the day. Although it was always in the back of my mind diabetes took a back seat to education itself. Extracurricular highlights included a one semester assignment with the school newspaper. Being the sports fan that I am my English teacher assigned me as a sports reporter. I enjoyed this for the reason that

I could not participate in sports physically yet I enjoyed them immensely as a spectator. This offered me the opportunity to interview coaches, players, and cheerleaders on occasion. They would be featured in stories written by our staff. It was a nice way to give the student body attention. This is a trait that many high school students enjoyed. We just started that trend a little earlier! One of the favorite features I was involved with was a humor piece that appeared in each month's paper. Its title was either "dream guy" or "dream girl." In it we would describe features from specific people within the student body and create a character by molding them into one.

When it came to trying out for sports I had to test the theory that I could not do so. One year I attempted to try out for our boys' basketball team. Had I been successful it would have been the equivalent of making the junior varsity team in high school. Although the first day of practice was strenuous I made it through. The second day however proved my undoing. The calisthenics portion was enough but the killer was when I was asked to run a mile first forward, then backward. Having failed to do that it was only a matter of time before the first cut came. This setback did not keep me from performing in physical education class. Over these three years coaches began to develop a level of respect for me that I probably did not deserve. I was made leader of several athletic teams. These were not competitive in nature except for the fact that we were competing against fellow students within the class. I feel the coaches did this in an attempt to spare me the embarrassment of being the last person selected on any team. I thank them for it but I already knew this would happen and it wasn't really necessary. One rule change they made involved giving my team an extra out in any inning in which I appeared at the plate as a batter.

In the eighth grade I became a scorekeeper for the junior high school baseball team. I did not travel with them on the road but just having the experience was enough. Dances and celebrations were held during the evenings periodically. I went whether I had a date for these dances or not. As a sixth grader I was just thrilled to have any girl dance with me.

With respect to running for student government I won successful election to Student Council in the sixth grade as a representative. I must say that in the sixth grade I was not eligible to become an officer. It did not matter to me. Just being a part of student government was enough. You might say it was here where I earned my stripes and a passion for politics. I distinctly remember being invited to a statewide conference of student government leaders. We had two representatives from our grade level. This required a coin flip to determine which of us would attend the conference in Flagstaff. I won the coin flip. Though this was new ground for me I was not as excited as I otherwise might have been because I would be guarding against diabetes a good deal of the trip. Nonetheless I enjoyed the experience and I chalked it up as a good educational exercise.

This interest in politics, news and sports was augmented by a specific game that we played in our history classes. Approximately every two weeks our class would get together in a current events style competition against another history class. Each class had some very sharp cookies in it.   After questions were asked, answered and removed from the board, the stakes grew higher. Games were often decided in the Double Jeopardy round. We played those games more for class pride than anything else. Our teacher often provided donuts to the top three news junkies in the room for that particular week. On more than one occasion I was within that top three. To this day I remain a very active news watcher. I even follow it to the point that I need separate cable channels for sports, weather and news. I figured the information and knowledge would be my power in life.

In the summer I was able to participate in recreation classes sponsored by the city of Phoenix Parks and Recreation Department from June through early August. Not only did I have a swimming pool at home I also had the ability to use it day or night. These classes were a perfect combination of exercise, skill, and sports that was necessary to keep the boy healthy in his off-season. There were many days when I literally went to recreation classes as early as nine o'clock in the morning, returned in midafternoon, swam hard in the pool, including lapses and/or keep-away games with my siblings, ate dinner with the family and capped off the evening with a backyard whiffle ball game or a swim!

When it came time for high school it would be another brand-new environment. The high school I entered was brand-new in the district. The school was only in its second year of existence and had no senior class in its first year. With a nervous anticipation I remember pre-registering for classes. Among the things we picked up as students prior to the start of the fall semester were our PE uniforms. For the first time in memory physical education was much more like a military instruction camp than a class full of casual students in school clothes. We were actually required to dress in uniform as part of the class! We were also assigned lockers. These would be our living quarters for books, and gym clothes.

Having taken a quick visual tour of the school's interior I realized it was on two different floors. Fortunately students had their choices of either stairs or elevators to assist them in class transfers. The campus seemed to sprawl over three different buildings. One of which was our cafeteria. It consisted of an outdoor patio/picnic area with a soda machine and an indoor bi-level lunchroom. Students had their choice of hot, cold or brown-bag lunch from home. Finally we really had a choice of what we wanted to eat. I remain one of those who brought a sack lunch from home most of the time. It was much more economical that way. A second advantage was the fact that we had our own citrus in the backyard. We could bring our own fruit to lunch while other students could not. The other two main buildings were connected and broke off in an L shape. There was also a gymnasium at the far south end of campus.

Being on a campus that large filled me with instant rush of excitement. Although it did not occur to me until later my class would be the last of the original four to go through this campus. Another advantage was the fact that some teachers who taught me in junior high school accepted the promotional challenge of teaching in the new high school. I, along with the rest of my classmates, felt a bond between those teachers and us. This would certainly make our transition and assimilation much easier.

I must say that my favorite subject in high school happened to be history. I had a couple of great teachers in this area who seem to relish the fact that they were teaching. I can only hope that that fire still burns today in the modern teacher. I feel my most dreaded subject was the sciences. As long as we stayed in life science, such as biology, I was fine. On the other hand chemistry was a nightmare. I was much more interested in the demonstrations put on by the teacher within the lab than trying to figure out chemical equations on paper. There was also the four-year English requirement. I enjoy creative writing and English grammar classes but was not quite as fond of poetry. Senior year was much more fun with plays by Shakespeare and dramas of the macabre written by Edgar Allen Poe and others of his genre. Of course there was always physical education. Although the school only required three years of such I took it for all four years in an effort to stay healthy.

I again tried to run for office within student government in high school. By this time however the election became much more of a popularity contest than it had ever been before. I don't know why this was a surprise to me because we had all been together now as friends and classmates for four years. After this much time people begin to know each other a lot better than they might otherwise have. As these relationships develop so too do cliques. As each one develops you struggle to find out which one you fit into. I was not a jock because I did not play sports. I was not a brain because while bright enough I was not the brightest bulb in the room. I was not a stoner because I did not use or believe in the use of illegal drugs. I was not a flirt because I did not chase girls. I figured I was physically slow enough that they would catch me! Instead I considered myself a friend of everyone except the stoners. I got along well with almost all types of students. Making friends was never easier in my life than it was in those four years.

With regard to extracurricular activities, I was one of those fringe students involved in a little of everything, yet not at a real high level. Instead of being a part of homecoming royalty I helped my class design three of its four homecoming floats over my tenure. Instead of running for student government I often took up causes for people who were running for student government. Athletically speaking I was offered a very special gift by the athletic department.   To this day I may never get another chance like it. I was given the opportunity to do public address work for our varsity baseball team and our underclass football teams. I also helped keep statistics for all three levels of our basketball program. This gave me exposure to all sports without feeling the pressure of having to play. It also allowed me to focus on a hobby of mine and give a few of our underclass athletes the attention that they so richly deserve, but often do not get.

One of these public address adventures that I will never forget occurred during the spring of 1984. I had been doing public address work along with a dear friend of mine, John. He and I were also classmates from the fifth grade. Though he was an avid sports fan and could probably play he lacked the level of coordination necessary to perform the sports. That did not stop him either. He helped me as a scorekeeper during baseball broadcasts over the public address system. If the sport was football John was my spotter in the press box. It was he who was telling me who made the tackles on which plays. He and I did much more together off the field and still do to this day. I thank you sir for your continuing friendship. Friends like you are hard to find. I know not how much different my life would be without your presence in it. Regarding the story I was about to tell, it seems as though the original public address announcer assigned to our state playoff game took ill. Needing a replacement someone asked our coach if they knew of anyone who could fulfill the duty. I am thankful that person did. This was probably the largest single event that I have ever done public address work for. In my excitement for this I warmed up on the baseball field with a backup catcher who was a part of our baseball team! It is that warm-up I remember more than the announcing itself. I do remember that we lost the game that night. The score was close but the loss was expensive. Our starting third baseman suffered a broken nose. This was simply a contact play and our player took the short straw. I never heard such silence on a bus ride anywhere in my young life. Instead of partying like most athletes might their concerns were toward their injured teammate first. This was perhaps one of the best displays of sportsmanship a young boy will ever see. I know a few of these teammates quite well. At least four of them were drafted into the major leagues at some level. All of them failed to make the big club. Two of the four went on to play collegiate baseball at Arizona State University. I had the pleasure of watching them both in their later two years. One of them is now the head baseball coach at Arizona State University. Though I don't follow the gentlemen day-to-day I can assure you that these people are class individuals and learned a lot of what they knew during their grade school years on the field and off.

One of the advantages I had in high school was the fact that I had earned enough credits as a student to take a reduced academic load as a senior. This allowed me to take on my first job in the summer of 1984. I was able to work full-time during the summer and part-time during my senior year. I started on an invitation from my father. He saw that I was bored at home and needed something to do. It was his idea to begin volunteering his son within different departments of Scottsdale, Arizona city government. I tried my hand in several different departments performing odd errands. Whenever times were slow, I monkeyed around with a type writer trying to perfect a slower than normal typing speed. While this did not come at a bad time, it was right at the edge of the beginning of the computer revolution. By the end of that summer, nearly every city department communication system was set up via computer!

The last of these departments I volunteered for turned out to be my home. This was the city's building permits department. The new environment and reception were priceless. Seriously, how many 17-year-olds do you know whose first job was that of a public servant in a government office building? Aside from getting used to the environment I was finding out more about the people who worked there. I was immediately enthralled by the number of activities, errands, deadlines. Employees had to deal with on a daily basis. It took me awhile to learn these people's names and associate them with the faces they belonged to. As time passed, I learned who was who, as well as some valuable skills. To this day, without them I don't know what the rest of my professional career might have been. I learned skills of filing, replicating on large-scale copiers and creating ammonia or sepia blue-line prints. All of this was great, but would eventually lead to a taste of the real world that I was not ready for.

My senior year of high school could not have been drawn up any better. In addition to only one half day of classes and the other half work, my favorite baseball team suddenly found a version of magic that could not ever be duplicated in modern times. Forty games into the regular season, they had already won thirty-five of them. By May, folks in Detroit were already thinking about playoff tickets. Over that first summer and into the fall, my comfort with the job I had continued to grow. The future looked brighter than ever.

Early on in the school year, an assistant principal stopped by to talk with me during lunch hour. He talked to me of how he had seen me in school and read about me in the administrative files. Usually, this means trouble for a high school student. In this case, nothing could have been further from the truth. This gentleman's concern was over a disease that was threatening me as well as him. The following were the exact words he used:" Jim, I have read quite a bit about you lately, and I've come to find out that reading is not enough. I know that you are suffering with diabetes. I want you to know that it may be too late for me, but if you keep fighting, and hang in there like you do now, medicine will catch up and you will have a cure for diabetes in your lifetime."

I was shocked by what the man said. I heard it before from many other people in different words. My initial reaction was not to believe it. I took this information with a grain of salt because my mother had been suffering with diabetes my whole life. Continuing on through the academic year, I went on to have one of the best personal years of my life. Who needed senioritis when you have a baseball team that was going to run away and hide from the rest of the competition? Who needed senioritis when you had the kind of job that most of your peers would kill for at their ages? The only things on my mind at that time were completing my senior year of high school, working hard over the summer to save money for college tuition, and spending the month of October celebrating a

Detroit Tiger world championship! I even kept a promise to myself that I had made four years earlier-buying only one yearbook to mark my high school experience. This one would have my picture in it in color. This is a privilege given to seniors annually. In those days, the price of a high school yearbook was approximately $35 per copy. I thought that at those prices, I did not need all four yearbooks, but only the one for my senior year as a memento.

In June of 1985, I graduated high school. This is no major accomplishment for most people. For me it was a major step forward in a plan that included a college education. I will never forget that night. It was unusually hot and there was talk of either moving the graduation ceremony indoors, or making it short in order to save the health of our parents, and fellow graduates. Our principal was unusually strict during rehearsals. We were often threatened with the fact that if anything was out of line during the ceremony, the whole thing would be canceled immediately and the privileges we got would not be granted to our peers. I did not laugh at the time, but I knew he was joking. There was no way that the ceremony would be canceled regardless of circumstance. What leaders we have in this class! Among them, a two-term student body president who is more than just a government leader. This guy was special! He not only led governmentally, but academically as well. A class valedictorian, this young man's life was nearly cut short by a drunk driver. In this accident, he lost his father. Our entire school bonded around this man and his family in his time of need. This crisis made our school stronger in its own way. As far as women go, we've had our share of those too! Erika had it all. She was smart, attractive, a student government leader, national honor Society member, cheerleader, peer counselor, and friend. Despite all of these things, Erika was one of those who did not allow her good works and smarts to tell us who she was. We should all be as fortunate in this area. She was just one of us.

As the ceremonies wore on, my nerves began to become frazzled. Any 18-year-old man who doesn't feel nervous like this at this time might not have a pulse. Although my dad considers this bragging, I will never forget the reaction when my name was called. I remember turning toward my parents and family in the stands and seeing nearly half of my peers on their feet, giving me a standing ovation. I did not feel worthy and to this day I still don't. All I did was accomplish what everyone else had to do! My best was yet to come.

As far as the job went, all things were go, except for one major developmental milestone that had yet to be accomplished. I did not go into any detail about my high school years regarding dates to either dances or proms. This is largely because I could not drive in high school, and my friendships were largely limited to those on campus. I did get out to dances and such things, but never dated. Had that experience happened, the next three years of my life may not have been such a social nightmare.

I turned my focus squarely into work for the time being. This was intended as a college tuition savings drive. I intend to work as many hours as allowed, and save as much money as humanly possible for any eventual college education. I did not know just how powerfully Cupid's arrow would pierce my heart. I have to say it here. Without it, I don't know what I might've missed. I do know, however, I would have saved myself a ton of heartache. I put my focus into employment, and the beginning of college education from the summer of 1985 forward. At the time, I had more than one objective. These were to put away enough money over a short period of time in order to start full-time college education down the road, and to gain a larger measure of respect from peers within the work community. I guess you might say that the results of this would be a mixed bag.

Both of these objectives were achieved, but there was also a lot of delayed pain. By delayed, I mean that most human beings accomplish developmental milestones at about the same age. In my case, they came much too late. The biggest of these was my first serious teenage crush. Most often, these happened at about the age of 15. Mine hit at about the age of 18. Furthermore, the setting for it should never have been in the office. Rather than go into details about what happened when and who said what, I will leave it alone. For me to go into it in any great detail defeats the purpose of forgetting about it in the first place.

This stretch was not a total loss. In the fall of 1985, I was awarded as most outstanding high school student by the Phoenix Mayor's Committee on Employment of Persons with Disability. The award itself probably was not deserved. I remember at least two occasions where I turned it down with my parents. I was too busy to actually go to a dinner like this when I had college coursework in front of me. At the time, my parents were trying to keep the surprise from me. My principal at the high school told them what was happening. This made attendance mandatory. The reward was $1,000 worth of valuable tuition money. At the time I did not consider it a scholarship for the sole reason that I figured you had to be super smart to earn such an award.

As I finished out the three-year period with Scottsdale, I was less than happy to leave. I did realize however, that life could not continue as it was without me being protected in one way or another by a future that lie ahead with better benefits.

Dramatic Years of Growth: 1987-1995

I began my foray into college education on a full-time basis in August of that year. I can remember being one frightened newcomer. This place was like a city within itself. There were some 40,000 students also seeking their place in the world! On top of this, quite a few of them were living in university housing. For the first time in my life, I was living entirely free of my parents. The first year of my existence was spent in a 13th floor small bedroom with a study area, restroom with a shower and sink. The only view to the outside world was a single window of dual-pane glass approximately 150 feet above ground. One of the biggest safety concerns I had at the time was for a fire when the elevators would not work. Many times a jerk would set off the fire alarm deliberately or while in a drunken stupor. This left me with a thirteen flight descent down the stairs to the outside exit. I can't begin to tell you how many times I had to make this journey between midnight and 3 AM.

Outside of that I was excited about what lay ahead. It was a chance to get away from 9-to-5 work. It was also an opportunity to finish up a dream that resided not too far below the surface of my mind. I had wanted to accomplish this since approximately the seventh grade. I did not know that the skills I obtained in high school regarding how to study would be severely tested. I began what was called an academic yo-yo-where you are on academic probation at the end of the Fall semester, but miraculously pull yourself out of the gutter by the end of the Spring semester. The only problem I had came in college algebra class. I had to repeat this four times! It was not for not understanding the material presented, but not understanding who presented the material. After the first year of this adventure, my funds ran low. I soon hooked up with the Arizona Division of Developmental Disabilities. These people presented me with the opportunity to finish my education.

By finish, I mean attaining a bachelor's degree. There were a few proposals within this opportunity, which change the course of how I studied, as well as how fast. The State of Arizona paid for my tuition and books. They would not pay for room and board. They would also not allow me to take semester breaks as normal. I would be involved in summer school for the rest of that career, as well as any graduate work that I might choose to do later on without their backing. The plan was to keep me going to school continuously until I had attained a degree that I was seeking. If government is going to pay for a service for you, they will tell you the terms you will abide by. Lesson number one: consider the source. This meant 12 to 15 hour course loads continuously until the degree was attained. Among the highlights of this reward was attending summer school for the first time in 122° heat. Students went to class in rooms that in some cases, lacked air-conditioning. After 13 years of living in the desert, I knew that this meant plenty of water had to be carried with you everywhere you went. I can remember sitting in a class with a beach towel wrapped around my head. Inside this beach towel sat a 3 pound bag of ice which would melt in order to keep me cool. In those days the swimming pool at home never felt more inviting.

A second and perhaps better benefit was a series of appointments with a psychologist. This turned my attitude and approach to studying completely around. The man did this with one of the simplest phrases in the English language. "so what." When I started to apply this technique, test scores themselves did not matter anymore. The material is what mattered! I decided to make an investment right away that proved priceless to me during my education. Anyone that has ever taken college courses knows the marathon early. Periodically you will be stuck in lectures lasting 3 to 4 hours each. To this day, I don't know how college students can stay awake that long without some store interference, which supplied either caffeine or sugar. I preferred the former of the two, along with a microcassette recorder. This tool gave me the opportunity to listen intently to any and all lectures, whether I was awake for the whole thing or fell asleep in class and decided to catch the rerun later before I went to bed. You might ask:" Why would I fall asleep in class?" The answer would be quite simple. Number one, the heat, was already stated above. Number two, I'm a diabetic. Over the years, one of the things I suffered with greatly was a feeling of being tired. Often, I did not know whether that meant the blood sugar was too high or too low as this could be a sign of either symptom. We handled this half of the problem rather quickly. The second half of the issue had to do with the devastating loss of self-esteem related to relationships with members of the opposite sex. The man's answer should not have surprised me. Still, it took a while for me to understand. I believe his exact words were to the effect that some people are going to like your action, and others are not going to like it. Either way, your action is your action and you should not have to change that for anybody. If you do, it is your fault. The reason you are changing your action is your responsibility. Since then, I have decided to leave women out of my picture. I have enough trouble with what is on the plate I already carry. The last thing I want to do is carry someone else's plate with me! I have to think of this the same way for the opposite sex. They have enough on their plate! Why would they want to care for me?

The new philosophy worked well for the here and now. It did little to alleviate the memories of years past. Neither did what I tried to do for my 21st birthday in college. Most college students bar hop when they reach 21 years of age. People offered me the opportunity to do the same in 1988. Instead, after approximately 4 weeks of training in my own neighborhood on a semester break, I walked the approximate 5.5 mile hike from my dormitory room to the office of my former employer. Though the visit was nice, it lacked the original feel it had as an employee. At the end of the day, the only thing that was still hurting was my pride. Physically, my legs and feet could say the same. I can distinctly remember the words of my nephrologist shortly after that trip. When I told him what I had done, he calmly said: "Are you nuts? For the sake of your own health don't ever do that again. I don't care if you did train for over a month in advance. The state of your health right now is not the greatest and your running days ended a long time ago." Lesson two, before one goes out on any strenuous exercise of any length, he should consult with his physician.

From this semester forward my academic performance began to improve dramatically. Grades which used to be average at best began a rapid ascent. Soon, A's and B's were a common occurrence. One more major hurdle remained. It became time to declare a major. Having tried business school, my real passion included mass communications. Journalism or broadcasting would've served me perfectly. When I toured different booths representing different colleges, another version of life's ugly truth was given to me. A representative of the public programs college told me that broadcast journalism or radio broadcast would not be the best place for me for the simple fact that 90 to 95% of the personality's livelihood was based on public appearance. This is not to say that I was not attractive physically, but a polite way of saying that physically handicapped people did not present a good public image. Tell me something I don't know! Disappointed, I fell back upon a field of expertise that I thought I might have. Up until then I had spent my professional life helping others. I thought social work would give me that type of platform. I had to realize now that success was not about money, but passionate pursuit of the field you enjoy. If there was one thing I thought I needed to do, it was give-back for all of the people that have helped me through the years.

Another nice benefit came along the way. The focus of the education narrowed as did the class size. I would now be going to class with the same group of 30 to 50 people, regardless of the course. When you attain major, the classes are fed to you in a specific order. They must be taken in order, or you will not graduate. That allowed me to make better friends for the simple reason that they were all with me! Academically, this is a smaller group of professors that you will see more and more often throughout the remainder of your education. It also meant a more intense focus upon your grades. Rather than worrying about the material, I focused on material retention, including conversion of taped material to index cards. This, along with material from text, much of it written by the professors themselves, gave me the added spark necessary to finish my undergraduate career on the Dean's list for the final four semesters. I never thought I would achieve this kind of academic success! But at this point I had learned also that the bachelor's degree was not enough. My feeling was that I needed to break a tie. If I left Arizona State with a bachelor's degree alone, any other individual who was fully able-bodied would have an advantage on me. The edge would have to come in the form of a master's degree. This was not done as a power trip, but as something to help a potential employer look beyond what he sees on the outside of the human body.

The enthusiasm was there. The attitude from my professors was not. I went to the man whom I had as a professor on different occasions and as an academic advisor throughout my education. When I expressed my intention, he implied that he did not think I had the wherewithal to survive a graduate level program. He indicated that he was afraid of my physical health. I understand the concern, but it frustrated me. Where was this fear while he was my professor in undergraduate coursework? What made him think that I would give him entrance fees to graduate school and then not finish the job? I had the choice to either take both years full-time, or split it over three years. The first of which would be a part-time year. The last two years would be full-time. I chose the latter option. I firmly believe that this was the right move because when I start something, I don't know when I will finish it. I just know that it will be finished.

On a positive note, I met a new friend. Mike is also disabled. This is an invisible disability to the naked eye. Mike is learning disabled. Among other things, he must slow his pace of reading in order to comprehend material. He taught me to play Pinochle. We would play when we were not studying. I remember many games which lasted through the wee hours of the morning. The only reason they did was pure enjoyment.

Mike and I both have an intense love of professional sports. Our friendship is so strong that I took part in his wedding and I recently hosted Mike, his wife Jennie, and his son Shawn for my 45th birthday. As a gift, they gave me the next two years of Major League Baseball broadcasts over the internet! This will allow me to catch any game from the voice of any team in the majors.

Mike married a woman who is also afflicted with Cerebral Palsy. She walks normally, but was left blind. The two of them have a four-year-old son who is now experiencing possible developmental delay. They each work within the San Jose' School District in separate schools. Mike teaches children in Special Education with a purpose of eventual transition into mainstream classes. Jennie works as an Educational Psychologist for the same children! Imagine having to work the jobs you love all day and bringing it home to raise your child! This couple has other dreams for themselves. Go get these dreams, guys! You not only shine for your community, but your country and society as a whole!

Good academic performance continued. In graduate school, it is not possible to receive a "C" grade without having to repeat a course. On a four-point scale, I managed a grade point average of 3.40. Despite this, I was one of only five students in the program to have to go to an oral board to receive a graduate degree that year. Upset by it, I began to trap myself in my dormitory room or any other place where I might otherwise be able. Fortunately, I saved the recorded lectures from the past 3 years. As others were buying their caps and gowns, I was fighting for that second degree. The morning of the exam came faster than I thought possible. I walked into the lobby of the School of Social Work with a shirt pocket full of information. Repeatedly, I would pore over the information written. A while before the exam, one professor came out and asked me if it wouldn't be better to return to my room, take a nap and return for the exam. I stated to him that he had already taken my money and inflated my grade to match. I would not walk away without a degree. Furthermore, I would not be back to repeat a course based on the fact that I made a bad mistake on one exam. I did leave, but it was for lunch. Having refueled, I returned to the site of the oral board. To tell you I was not nervous would be the largest lie I could tell. With relief, I passed the exam. As I left the college, I wonder how I did not cry! Since alcoholic beverages were a no-no, I celebrated with the purchase of two bottles of Martinelli's sparkling cider. I decided that Mike and his suite mates should be a part of any celebration that I do because they were there at a time when I really could use a friend or four. After a hard-fought battle my parents rewarded me with a trip to Hawaii with them. The things we did there will always stay fresh in my mind.

Lava Rock on the "Big Island"

Triumph With a Personal Price: 1991-2000

There was a personal price paid in the previous section. Because of this, I left out one of the most important placements in my academic career. Although it was for college credit, the events within it did not take place in the classroom. It also had nothing to do with academic progress. In this early stage of my life, I was quickly able to learn one of the most important features of human existence.

That first field placement year was in the Spring of 1991. We call it Spring academically because the spring semester begins in January. My field assignment was within the field of medical social work. I don't know how it happened that way, but the faculty and staff in the Arizona State University School of Social Work must have been watching out for me in more ways than one.

Upon arrival in a dialysis clinic, the first thing most people notice is an overwhelming sense of refrigeration. Arizonans use air conditioning to survive triple digit temperatures in the summer. Dialysis clinics use air conditioning as a means of reducing and/or eliminating bacteria. The average clinic maintains a thermostat temperature of approximately 65°. Compare this to a home temperature of about 74°. Once one becomes accustomed to the cold he will be able to focus on the job at hand.

I suppose here is as good a place as any to explain what the dialysis process is. This procedure is used in place of a body's kidneys. When they fail, it is the process of dialysis which will cleanse the body of toxins built up in daily living. When kidney function decreases; the body's ability to eliminate liquid waste decreases. When kidney function drops below 15%, dialysis becomes necessary to aid the body in eliminating its own waste! Unfortunately, the machine is unable to eliminate solid waste. Instead, it pulls liquid waste. The result is dramatic weight loss in a very short time. Dialysis patients have to follow a restricted diet. First, it forces patients to retain as little fluid as possible. Also, regular nutrition and medications to combat chemical content within meals becomes necessary. The only way to stop dialysis is to receive a kidney transplant. In my year with this clinic, I saw six patients leave in body bags! I did not see this as a field placement, but as a possible final chapter in my own life!

Early on, I can remember introduction and a sit in with medical staff who managed patients in that clinic. During the course of this meeting, the social worker in charge would review patient status on an individual basis. This gave me a secondhand opportunity to listen in on a patient's struggle. It is largely because of this field placement that I have decided to take the approach I do. There will be more about that shortly. During the course of one of the early conversations came a very painful lesson. When you're introduced as a diabetic, do not volunteer the fact that you have not seen a doctor in at least three years! I am glad that I did, but what follows is something I would not wish on any person.

I did follow this doctor's advice. I went to visit the nephrologist that my mother had for several years. At the time, I did not have any idea just how much damage diabetes had done to me. This doctor took my blood pressure and read it back as 170/110! The next step was self-explanatory. For the first time in my life, I would be placed on a regular blood pressure medication. I was 20 years old when I went to school for the first time. By the time I turned 23, battling diabetes turned into a brand-new ballgame!

As long as I was out taking care of myself, I thought it would be important to have my vision checked. Blindness is often a complication of diabetes. At the eye exam, the doctor told me that within 10 days I would lose sight in my left eye if I did nothing! My objective was clear: revisit this doctor within the 10 days and have some swollen blood vessels which had leaked repaired by laser surgery! At that time, eyeglasses were not necessary. Today they are.

Having gone through this dramatic calling out process, I returned to the field placement. The result was a much more focused attention to my own disease as well as an understanding of that many of the patients in that clinic were in dialysis because of diabetes. All that was left was learning the finer points of medical social work. Here, I am speaking of aligning with community resources. I often gave patients the ability to bring medical bills, utility bills, nutrition issues requiring emergency food boxes and more to my attention. I also was taught office procedures to handle while my supervisor was away at another clinic upstate. Through this, I had occasion to speak with patients directly. I did not focus on what I knew about the disease but rather on what they knew about the disease. As an intern, I had no authority to counsel patients on anything. I had been trained to use the skills learned having to do with handling community resources. It was here that I learned how to handle a Medicare explanation of benefits form. As a patient using it now, I understand what these things are.

As I said before, these were the reasons the field placement was so valuable to me. I now had a preview of what the future could hold for me, and I dreaded it. I have already covered much of what happened over the next four years. So let me proceed from here to the summer of 1995. Despite all I had accomplished, I had still not learned to drive an automobile. I can't remember whether this was a graduation gift, or not, but I would not have minded. I had an instructor whose specialty was teaching the privilege of driving to disabled drivers. What a thrill! As I slowly picked up that skill, my overall confidence as a human being skyrocketed! I had my confidence shaken like any other teenage driver, only this time it was not my mother or father doing the teaching. After approximately 3 months of learning, I received a license to drive. During this period, I was able to go out with my parents and purchase a vehicle from a previous owner. It was done this way in order to avoid the pressure of a car lot.

It just so happened that very weekend of the car purchase was the same weekend of my 10 year high school reunion. This was a three day weekend of events to celebrate with classmates from your high school the events of your life over the last 10 years. Most of these people were married, had children, been divorced, or even saw the world. Having just learned how to drive, I could not think of anything better to bring to a high school reunion full of students who rode home from school with me. I was simply a dropping off point along the way. I was able to drive to the reunion. More than anything, this is what my high school peers were celebrating. I will never forget that three day weekend. I quickly learned that after high school, reality sets in for all of us. There is no longer that sense of "one-ups-man-ship" that happens so often in life today. Instead of being jealous about what others had or didn't have, this was a weekend to celebrate what others had done. It was a chance to look at one person or another and find out what made them successful. The time for jealousy or envy is when you're sitting at home, away from them. When one looks at his high school reunion guestbook it becomes clear that we all come from different backgrounds, different income levels, different races, different cultures and so forth. The only thing individuals have to fear is not trying at all.

In July of 1997, one of my worst medical emergencies ever occurred while behind the wheel. I had just come home from work. Along the route, I noticed a feeling of tiredness. It was an extremely hot Arizona afternoon. I could barely remember getting through the front door of my home at the time. I was exhausted enough that I simply lay back in the chair and tried to take a nap. Supper eventually arrived without incident. After finishing evening dishes, I proceeded to begin work on my academic studies. Just before that, I remembered I had a little laundry, which needed cleaning. We had a large family, which meant that I had to switch other loads from the washer to the dryer before I could even start mine on many occasions. As I was making one of the switches, I passed out in the laundry room crashing to the linoleum floor! I don't remember the rest of it. All I know is what my mother recalled. My family had to rush me by ambulance to a hospital. I had been suffering seizure activity in my brain for some time. I was comatose then, and over the next four days. I remained in one which was drug-induced. As families do, mine gathered around my bedside and outside of the room for those days. As most families do now, they tried to bring me things that would stimulate my brain. Simple conversation, Walkman cassette players with your favorite music in them, complete with headsets, and so forth. Four days later, I came out of that coma. I remember my response to mom and my awakening." I feel like I am in Smitty's in a makeshift hospital."(Smitty's was a Valley grocery chain which had a frozen food section) To others, this comment may have had no focus. It shouldn't have! Anyone who is emerging from a coma may and often does suffer a lapse of memory. Over the next week I began to recover my senses, faculties, and ability to walk. I remember being told at the time by my doctor that he did not know how I managed to avoid death. He also told me that he had no idea exactly when I would be going to dialysis for treatment.   He told me to just remember. Maybe six months from now, a year from now or five years from now, but dialysis is next. I was placed on anticonvulsant drugs for the first time in my life.

Over the next two weeks at home, I took all medications as prescribed. During this period, I noticed a gradual decline in my motor function. People had always joked with me about the fact that I appeared to be walking like a drunken man. For the first time, I felt like I was walking as a drunken man. I decided to read the side effects of this new medication. Primary in those new side effects was the precaution of a slowing nervous system. Without delay, my family and I decided to make an appointment with the neurologist who had been seeing me during that hospitalization. He immediately changed the anticonvulsant medication to another form. This was my first known allergic reaction to a medication. I proceeded to take that drug for the next 14 years of my life. To my recollection, I have only had one other seizure in my life. It was much smaller than the original and easily spotted. Right now, I do not take any anticonvulsant medication. I have been seizure free for so long that I have decided the need to take these pills is no longer necessary. My neurologist agreed, with the following stipulation: should I ever have another seizure, the old ghost of anticonvulsant medication will return and it will stay with me for the rest of my life

Approximately six weeks later, I had a follow-up appointment with my nephrologist. It was my mother's birthday. Other than that, I felt as if everything was normal. Once I stepped on the doctor's scale, I knew something was wrong. It seems I had gained roughly 7 pounds over a seven-day period. In my mind and heart I knew this could not be a good sign. When the doctor told me the news, I was devastated. I can honestly tell you that I have never cried in a doctor's office in my life until that day. It appeared as though my final chapter would start. I asked him if I could spend enough time to take my mother out to dinner for her birthday. He asked my father two things for me in response. First, he told me that I could not go to dinner with my mother. I had to check into the hospital. Second, he appealed to my athletic side by telling me that he thought I had more at-bats left. He asked my father if he would go through a workup to determine his ability to donate a kidney to me in a transplant. I don't know about the rest of my fellow readers when I say this, but when I have a family doctor who cared enough about me to ask my own family about help such as this, I know that I have a friend as well as an ally on my team! Doc, wherever you are, if you ever get your hands on this document, please share it with others. Although you're no longer in practice, the message in this book will be clear.

Over the next five days, I was in dialysis on a daily basis. In that week, I went from a stacked 185 pound man to a week 130 pound man! Along the way, the doctor would come to me when I was about to order a meal. According to what he asked me, I knew what to order for food. I resented it at the time, but once I understood why this had to be, I got over the resentment fairly quickly. Over the next three months, I transitioned into my own chair in a dialysis unit. I was now living as my clients had lived for several years earlier! I can remember one instance where I told a dialysis nurse, I was fine at the end of the treatment. Four steps later, I was on the floor! There are days when your blood pressure drops so quickly as a result of changing from an upright position to an upstanding position that balance is a dream. As I adjusted to dialysis, these things disappeared.

In the meantime, my father participated in and passed his workup in order to participate in a kidney transplant as a donor. In January of 1998, surgery was initially scheduled. Just a day or so before surgery, the hospital called and delayed it. I was initially upset but I realized that a sick surgeon is no surgeon to do a transplant on a human being. One week later, the reality happened. I do remember the morning of that surgery a blood sugar level above 400. This is because diabetes had taken a toll to the point where I could not eat without seeing a tremendous spike in blood sugar levels.   Rather than immediately perform the transplant, my father performed his half first. While he was doing this, I was in dialysis for two hours as a preparation for my half. When he arrived out of surgery, I did not know just how well he was doing. As I was injected with IV muscle relaxants the effect of these things brought tears to my eyes. It wasn't a drug that brought tears to the eyes, as much as it was a feeling that my father had done something heroic. It is this writer's opinion that short of going to war for his country a father can do nothing more than donate part of himself to his own son!

I came to a realization after surgery that this was not the end of the road. Kidney transplants only extend the life of a diabetic. They do not cure the disease itself. I must keep vigilant regarding diabetic care if I want to avoid a quick return to the dialysis unit. What came next was a recovery and an eventual job within the Building Services Department for the town of Queen Creek, Arizona. By and large, this job worked much like the first one in Scottsdale. The only differences were in the lack of any interest in a girlfriend, and a new skill or two picked up along the way, which allowed me to process, charge customers for and electronically file building permits.   This gave me a little bit more financial responsibility. As the years progressed, I knew that I felt at home. I could've stayed there and I might have to this day had it not been for another youthful mistake. During a particularly stressful day I blurted out to a co-worker that I might "Go-Postal" on her. She took this slip of the tongue as a threat and by five o'clock the next day, my career as a Building Permit Technician was over. This was not said in front of a customer. It does not matter where it is said. The choice was given to me to either resign or be fired. Devastated, the only choice I had was resignation. I spent the next eight months of my life unemployed continually looking through the newspaper want ads for openings. Whether by conventional methods or applying online, efforts were of no avail until November of 2000 when a big break came my way.

The Chase Years: 2001-2005

The first part of any new career is nailing the job itself. Although I had worked in an office before it marked the first time in my professional life that I have ever worked in a high-rise building. In addition to the challenges of the new environment I arrived in a wheelchair. This presented a brand-new obstacle I had to overcome in obtaining employment. As before my attitude was simply that the job I was after would be mine for the taking. It has never served me well going into an interview with a negative attitude. Once I rolled into this high-rise building I was spellbound by the cubicle environment that I would be in for the next several years.

As the usual interview questions were asked I quickly fell in love with the environment. Although it was a cubicle it would be my cubicle. I seemed to satisfy the interviewers with my answers. I still had questions related to my handicap and the employer's ability to look beyond what they see in a wheelchair and instead at a gentleman's performance. Having stressed that in the interview probably sealed my hiring. Among the fringe benefits of this job was the fact that I could set my own hours. This is of particular importance because I needed several doctor visits to maintain health during this period. Having an open Friday to do this without spending company time meant that for the most part I could take care of physical needs without costing the company money! I was also in a vibrant college community. This meant there were plenty of nightlife activities for me after work, such as movie theaters, bookstores, bars and apartments in which to live. Additionally

Arizona State University was located within four blocks of the office itself. This afforded me the opportunity to use Arizona State University's Recreation Center after hours in order to maintain physical strength. As time passed, and the stresses of the job became more severe, I was able to use the gym and receive full body massage about once every two weeks in order to relieve stress built up over the pay period. That late November afternoon opened up a lot of doors for me. The official offer for employment letter stated that the job would not start until January 1 of the following year. This gave me roughly six weeks to update my business wardrobe, look for and obtain an apartment for rent in the area, and prepare in general for the next five years of my life.

When training began in January my initial reaction was one of drowning in fear. My trainer put that to rest in a hurry. I became comfortable with his teaching style and the language he had to use in order to read from the policy manual that we had to absorb. As I adjusted the competency level increased. Everything seemed to be flowing smoothly until the first day we went to the floor. In a sense this was expanded training. I was now learning from people who already had jobs within the center. My chief responsibility was one of the learning the tricks of the trade. I have already attained much of this knowledge in prior jobs. There is an art form that is played out when one deals with people on the phone as part of his job. Having done so for years prior I picked that side of it up quickly. The parts I struggled with involve speed of the phone call! Picture yourself if you will as a customer calling with a credit card issue which put your credit rating in jeopardy. As a customer you may not be as financially literate as an employee who is working with you at the time. Now imagine the fact that you have less than two minutes to handle an average phone call. As the job evolved this became the new expectation. Imagine having to do this at least 100 times per day five days per week. Is it any wonder that the average employee's nerves will be frazzled by the end of one week?

I absolutely loved working this job. It taught me a great deal about personal responsibility. I wanted my own credit card because that was one of the benefits of the Chase employee. Rather than immediately jumping on and getting one I waited until I had one full year in the business. I was able to listen to customers complain about issue after issue for that long a time period until I was comfortable with credit card policies. *I now had a very good idea of my monthly budget was and was able to use an automatic payment system where your credit card bill is paid on time every month by automatic withdrawal out of your checking account.* I cannot tell you how many times I offered this service to clients over the years only to be turned down. It was the same clients who would call back within sixty days of the original call with additional reasons for being late with a credit card payment, seeking waiver of a late charge, disputing charges they claim they did not make, requesting lower interest rates on their own cards and the like. The lessons learned in that first year only bolstered my ability to pay a bill on time. A simple rule of thumb I had and that I would offer to everyone is that if you don't have the money to pay for something don't purchase it until you do. As I got more comfortable with the system I earned airline flier miles with my card. I could do this very easily because month-to-month I had set expenses that were always going to be paid. Between rent, prescription drugs, groceries, and medical bills I was able to rack up approximately $1,100 per month in expenses. I was earning an average of $1,600 per month. You can see that this job was not making me rich but it was making me more than enough to get by as a single individual. Imagine putting 1,100 frequent-flier miles on a credit card every thirty days. Basically this means a free trip every year if I so choose. I did take advantage of a couple of such trips. One was to San Francisco, California, and there are at least two others that occurred over the next four years.

As I stated earlier there was always the trouble with beating the clock. Over time I tried every trick in the book that I knew in order to help me with that struggle. It got to a point where I was documenting the time elapsed after every phone call in days. This was helping me with the job itself but killing me in terms of stress level. As you know I was suffering from hypertension secondary to diabetes. This kind of high-level stress does not bode well for somebody fighting blood pressure issues. It also plays havoc with one's blood sugar and insulin schedules. I can still vividly remember conversations between staff about my future with the company. One such discussion ended with the comment: "Don't worry about it. With a record like that he won't be here very long." How I kept my cool after that conversation I'll never know. My immediate reaction was kept inside. I was thinking: *"what record? As far as I know I hadn't been sick. I was at work every day I was scheduled. I dressed the part. When available I worked emergency overtime."* Within two years of this outburst that gentleman was forced out of a job by downsizing. I had the misfortune of running into him on an elevator ride when he heard the news. I couldn't say a word to the man because I respected him greatly. But inside I wanted revenge. I had it right there in the elevator shaft. Who's not to be around long now? This was the question I continued to ask myself all the way down the elevator shaft.

One of the benefits we had with this job was the opportunity on a monthly basis to review our past month's performance. It seems as a whole that they were really happy with my performance on a daily basis. My biggest weakness was the battle of the clock. This battle was compounded after about nine months when the focus of the company changed from a customer service oriented job to a serve and upsell approach. In my eyes all it did was exploit my weaknesses. Now I not only had to talk fast but offer products and services for sale to clients. Let me tell you right now. A salesman I am not! It is hard enough for me to serve a customer with their immediate issue without having to sell the company agenda on top of it. As the supervisors changed my general opinion did not. For a brief while this entire argument took a different course in the spring of 2002.

After work one night I returned home to my apartment. Within about three hours I had received a phone call from my father. He told me that mom was very sick. She had suffered a massive heart attack after running dialysis earlier that day. The symptoms hit particularly hard upon leaving the unit. She was taken to emergency at the hospital because of the symptoms. Diabetes had already taken a toll on my mother's health. By the time she suffered this coronary she had already lost much of her eyesight, hearing, and the index finger off of her right hand. Over the next three days mom continued to cling to life. Each night Dad would give me calls updating the situation. On that Sunday I was up early preparing to go to work. I was in the shower at the time the phone rang. I wondered who would call me at that hour. All of my friends knew that I worked Sundays. Three such calls occurred while I was in that shower. I raced to pick up the third but was too late. I went back to finish getting dressed for work packing the medications for the day and such. Just as I was about to open the door to leave I heard a hard series of knocks on my door. When I opened that door the site of my father in the threshold of my apartment door told me the obvious. Death visited my mother that morning and I still had to go to work. This time the purpose was entirely different. This time I had to request bereavement leave before the shift even started! This is what a businessman does when his mother passes away. Just once I would've loved to be home with the family when it happened.

The trip north to Scottsdale Memorial Hospital Shea Campus was a battle of trying to maintain composure on one hand and getting as much brief information from my father on the other. I don't know how I managed to keep a straight front all the way up to the front of my mother's room. I saw my doctor's face and could take no more. It was all I could do to throw my arms around that doctor in tears. I told him he had done his best and that she was going to a better place. In mine and anybody else's eyes this lady had suffered enough. Upon seeing my mother a lot of that sadness turned to thanks and anger. I resent the anger part. I had no business being angry. Mom lived a full fifty years with diabetes. If there is one thing she did teach me it was how to fight this disease. At that point I made a promise to her that I am still keeping to this day. First I would take care of my diabetes. I have. Second I would cure the diabetes. I have. Third once this was done I would help others do the same.

It's hard to describe the feelings one has when his mother dies because they are so individual. I can't begin to describe the myriad of emotions. I went through with the family gathered with us over the next few days. I thought men were supposed to be the strong ones. That's what everyone was telling me. I have never been an emotionally strong individual. I think this is part of what disability has saddled me with. If I was physically stronger I could actually feel like a man. When you are disabled, such as I am, one of the areas hit hardest is emotional control. This trait has followed me throughout my entire adulthood. Most men have an innate ability to keep their cool. I have to learn this craft. Often I am forced to speak my mind in order to get a point across. Often that speech is given in anger. If I could only overturn this bad habit future endeavors in my life would be easier to handle.

Shortly upon return to work I decided to take up my own mission. I realized at that point it was either going to be Diabetes or me. Shortly in my case meant approximately eighteen months. During this time I was subject to extreme swings in blood sugar levels while on this job. When sugar levels were high I had to choose whether or not to even be at work. When they were low I at times passed out on the floor. This does not look good if it happens more than once. I can remember at least two conversations with company nurses after such incidents. These arose largely because of a severe switch in insulin types. I had graduated from U-100 insulin to a type of insulin that was faster acting than regular insulin. By faster acting I mean that it would take effect within twenty minutes of injection. This juggling act made watching what one eats paramount in a battle against diabetes. I can distinctly remember one such morning when I came in before work and checked my blood glucose reading. When it showed 629 I knew I had to go home and take the day off work. My new boss at the time understood and let me leave for the day to return the next one when I had it under better control.

In the fall of 2002, a woman by the name of Deb Harvey changed my life in more ways than one. She would become supervisor number five in my Chase career. Her management style and energy level was something I had missed as an employee for years. That addition alone checked my fire into gear. This lady was different than most managers. Rather than focus on a company's bottom line objective at all cost she focused on employee's abilities and used them to get the most good for the company. In addition she was a professional about it through and through. This is not to say that she was a stuffed shirt. Instead it is to say that this woman worked with the employees she had as if they were individuals in her own family. I eventually tried sales at Chase and was remarkably successful. For six consecutive months I brought home bonus checks based on the sales I made during the month. Ironically quality was not being affected because I would bring home bonus checks for that as well. I must say that in the best year of my Chase employment I made nearly $30,000. This is not to brag but compare it to $20,000 as a regular customer service rep. One-on-one performance reviews each month were in much better detail than others. I was actually discussing with her an opportunity to get away from the telephones. I did not want to die on a telephone! This financial services giant had so many opportunities available outside of the telephone cubicle that I was dying to get the one. She understood that message and tried to have me train after hours for other jobs. One such opportunity actually came within the mortgage department. I went into that interview with the same approach that I landed this job with approximately two years earlier. Too excited I failed the written portion of the interview miserably. Nonetheless I thank Chase for the opportunity given at the time.

I felt very much at home with Ms. Harvey's leadership. For once it was as if I was on the same wavelength as my supervisor. Conversations took on a much more personal meaning during our one-on-one sessions. When she asked me for details about the struggle I was having with diabetes and cerebral palsy at the same time I could tell her those details with confidence. To this day I can never remember putting a boss in tears. I did that day! All of this really took a step forward in 2003. I had just left on bereavement leave yet again due to the death of my father's mom. Grandma lived to be eighty-nine years of age. Just a month short of her 90th birthday she suffered a severe stroke while watching television at one of her sons' homes. I received this news much like I had my mother's passing and chose to go on an airplane trip for the funeral.

I can't remember how I felt that day except to say that it was much less emotional than that for my mother. In grandma's case she had lived a full life. That is not to say that any woman or man should die at any age but it is also to say that ninety years is a heck of a lot more than fifty-six. She was able to raise four sons. All four of these children have gone on to live wonderful lives of their own. The only times I remember crying were when I was by myself in another room of the house. When I look back at grandma's life I see the good things that happened. She was married just short of 50 years to the same man! How many marriages could stand the test of time today? She also lived in the same home for all of those fifty years! To see the house today is to see modern change right in front of your eyes. If you are in heaven tonight grandma rest in peace for you have earned it. Upon returning from that funeral I began to pursue a possible cure for diabetes through transplant. At the time pancreatic transplants were considered experimental. As far as I knew they would still be. Dr. Cherrill came through again, just as he had before. With his referral I eventually received his blessing to pursue such a transplant. The first thing I asked was to meet with someone within his practice who had gone through such a transplant. It just so happened that as I was coming out of that appointment one such patient was waiting to be seen by the doctor. When I asked her how long she had her pancreas I was shocked to learn that it already had been a decade. I asked her where she received it, gave her my thanks and left the office. It was this visit to the doctor that opened the door for what ultimately happened. I consulted my benefit coordinator at Chase to find out if pancreatic transplants were covered under their insurance plan. By the end of that Friday the answer came back in the affirmative. In addition to that the procedure was covered 100%! The emotion that I felt in that apartment room that afternoon cannot be put into words other than to say that the promise I made my mother regarding curing this disease was about to become reality.

Once I had proper statistical information I pursued a call through the University of Maryland Medical Center. At the time these professionals had the best track record in terms of procedures done and five-year survival rates. Once they had set an appointment for me I quickly called my sister Molly in Dumfries,

Virginia. She was about a 2 1/2 hour drive south of Baltimore itself. I had already visited her approximately one-year previously on a vacation to Washington, DC. When I told her what was at stake she gladly accepted me and my father's visit. The only struggle I really had was presenting this to my boss. I had no fear that she would allow me to take the leave but how would it be classified. Was it personal vacation, or family medical leave? As it turns out the answer should have been obvious. It is always considered vacation unless you are literally receiving medical treatment. I took it as such. In April of 2004, I took a trip to Baltimore.

This day turned out to be the most educational of my life. By day's end, I had fourteen tubes of blood drawn from me, went through another course in transplant education, a physical, and a psychological interview to determine whether I would be a proper candidate for such an operation. They interview patients because of the limited supply of available organs for transplant. They do not wish to transplant an organ into a diabetic for the simple reason of not wanting to take shots anymore! (Show me a diabetic who wants to take shots in the first place and I'll show you a masochist.) I gave them perhaps the best reason that I could possibly come up with under the circumstances. I told them that I had a job to return to. Performance wise this job was suffering because I had diabetes. I also felt that I could perform this job with far less stress if I was given the transplant. Without hesitation the doctor and his surgical staff agreed to proceed with surgery. The exciting part about it was the fact that I only needed a pancreas. Under normal conditions at that time the waitlist was about six to nine months long. Hearing this news I didn't know whether to faint, cheer, cry or just shut up. I chose the latter. The last thing I wanted to do was celebrate a short waitlist when there were others in that room that had to wait three to five years if not longer. Molly, my father and I spent the last part of that vacation touring Washington DC, the Baltimore inner harbor area and other historical sites. What a thrill it was to see our nation's capital as a tourist with a much higher objective. Returning to work I asked for and received time with my peers and my boss in a group to share the good news with them. I also wrote down lyrics to a very special song written by Don McLean. If anyone wants to they ought to listen to the words in the song "Crossroads". Just a couple of listens to that song will make you realize just how powerful four minutes behind a microphone can be. I have often used music as a vehicle to describe emotions that I felt personally.

Things were different when I came back. For the first time in my life I was anchored to a pager everywhere I went. If that pager goes off I may be asked to jump on an airplane and fly clear across the country to receive the pancreatic transplant. During that period I tried to put it out of my mind. It will get here when it gets here. During this period I returned to the gym regularly to stay in shape physically and to receive the massages which relieved physical and mental stress built up by the job. I wanted to be in the best condition possible upon arrival of the transplant.

Things were not always somber around Deb Harvey. There are often times of high stress. One such period arrived during a sales marathon. The object of that month was to have as many people as wanted to convert credit card debt to our company from another. Offering the balance transfer and selling it was my specialty. When this was performed it was imperative that we read disclaimers telling people that they must maintain their credit card payments on time lest the transfer terms be forfeited and converted to regular interest rates. Long story short we had one of these meetings. While we were in that huddle together as a team I suggested that one of us stand up and read the disclaimer regarding the payment schedule in the voice of Elmer Fudd. I did so in this meeting and left my supervisor laughing. Her face was beet red and she was pounding her head on the desk. It is an absolute joy to have a boss with that kind of sense of humor!

Later on she brought me a relative to visit. It seems as though Deb's 13-year-old niece was a diabetic. Unlike me she often went to school without taking her insulin shots. The result was that her blood sugar levels were way too high. I had to talk to her in front of Deb. This was not done as a father talking to his daughter but as a friend talks to another who is in his same boat. When she found out that I was wearing the beeper for the sole purpose of eliminating juvenile diabetes that I had been carrying for almost thirty years I hope I blew her mind. This operation was not available to anyone under the age of twenty-one and still isn't. If she has followed the advice I gave her that day she may be eligible for a pancreas transplant within the next three to five years.

In July of 2004, I had to make another change of supervisor. I did not find out in the conventional way. Rather I found out by a postcard in my employee mailbox. I was so busy during the day and most days that I could not read the mail before I started work.

I spent my lunch hour with it. When I read that I was changing supervisors yet again I could not understand how to break that back to Deb. This had nothing to do with the fact that I was changing supervisors itself but **why now?** When she met with me later in the afternoon I found out why. I did not go in there as a dummy. I went in with the belief that I was being sold out because I would not sell products unless I was asked by a customer to sell them. Her answer to me proved both shocking and positive. She had told me that I was valuable enough that in return for me she wanted two individuals who could sell. This proved my argument that to her she was in this job for money because her paycheck would fatten the more people she had on her team that would sell. I may not be an economics major but years of college education taught me the business side of business. I believe her exact words were: "Jim, you are getting a brand-new body. When this is done, you won't even give a damn about Chase. You will be off doing so many new different things that we will be the last thing on your mind." I thought about that for a moment and I wondered why in the world I would leave. This company gave me that cure on a silver platter! As such I would be forever in debt to them job or no job. I would definitely come back. Reluctantly I changed supervisory teams.

Under this new pair of supervisors things would be far different. They didn't even want to hear Deb's name! I thought this was extremely funny because almost everyone else in the company loved her by reputation and name alone. I tried to keep that in mind. Over the next few months the ravages of diabetes continued to take their toll. There was an incident at work where someone had seen me injecting insulin into my body within a restroom. Rather than confront me about it they went to my boss and ratted on me as if I had committed some kind of a felony for taking care of myself! I had to face up to that and explain to her in no uncertain terms why I was carrying the cases with me full of needles, insulin, and blood sugar testing equipment. All he had to do was ask me! I would have been glad to tell him why he was seeing what he did. I can't very well take it on the floor! One of the things I told that human resources representative was: "Do you want me to take the insulin in the restroom away from employees or do you want to see me in the hospital where your company pays the insurance bill because I didn't take the insulin?" They never asked me about it again.

In early September, 2004, while reporting for work on a Sunday morning the pager went off. When I checked it to find out who was calling I knew I had to make a beeline to the manager's office! This was the last thing I expected on Sunday morning. Having called the page back it was the University of Maryland and they did have a pancreas waiting. They did tell me at the time that there was only one such organ available and that they would not board an aircraft if they were me until they got a confirmation call from the hospital itself. I then went home and spent the rest of that Sunday waiting for a confirmation call that never came. I figured that someone else had received that pancreas and that my time would come. Disappointed I went back to work Monday. Then approximately three weeks later on a Sunday morning the pager rang about 5 AM Arizona time. This time I was informed that there was another transplant available for me. This time the only choice available was to call my boss at work and explain to her what may happen later in the day and that I would not be in to work. I also called Deb Harvey who had a Sunday morning shift. When I told her the news she acted as any leader would. She wished me luck the whole way saying that she knew how long we waited together for this moment to arrive and although I was not part of her team anymore it still meant the same as if I was. Eagerly I boarded a flight to Baltimore at about 11 o'clock in the morning. When I landed in Baltimore that evening shortly after 8 PM I called the University of Maryland Medical Center back only to find that the organ they had received was not viable for me. They had another person who matched the criteria for that organ better than I. Devastated I went back to the Southwest Airlines ticket counter and purchased my return flight to Phoenix. Out of the kindness of their hearts they also gave me a return ticket for when the transplant actually happened. If there was ever a time in a book for a charitable giving plug this is it. I wish to thank Southwest Airlines personally for their gift. I guess it is part of Southwest Airlines' corporate mission to be charitable in times of crisis when one has made the initial effort to handle the crisis himself.

That long redeye flight back from Baltimore to Phoenix could not have been sadder for this man. I don't know how I contained tears all the way back to Phoenix. All I know is that when I arrived back in Phoenix. It was already pushing midnight. By the time I returned to my apartment it was nearly 1 o'clock in the morning! Work resumed the next day at eight. Rather than taking a day off to recover I tried to put up a false front by going to work because I figured I would've been there anyway had the pager not gone off.

This was easier said than done. As soon as I reached the double doors at the ground floor to enter the office building there stood both of my supervisors asking me what I was doing there. I could have given them three guesses and the first two would have been incorrect. Instead I explained to them what happened and tried to keep up a false front. After about 15 minutes of preparation I could take no more. I quickly went to my manager's office and explained the details to her. Her response surprised me but it shouldn't have. She said, "Jim, what you have just gone through is comparable to that of losing a loved one. Frankly, I find it crazy that you would even show up to work today." I want you to go home and take as much time as you need until you get your head straight. When you think you're ready, come back." Better words were never spoken. Over the next four or five days there was a perfect dose of therapy. That weekend I participated in the Juvenile Diabetes Research Foundation's walk to cure Diabetes. This was an event that took place about one-mile from the office. It was a sea of some seven to ten-thousand walkers who either were diabetic or knew someone who was. Many people wore shirts memorializing family members and friends who had passed away from the disease. They were out raising funds to cure this killer. That 5K walk did more to cure the depression of that flight than anything imaginable. My attitude then became one of a diabetic this year but a cured diabetic next year when I returned to the walk. At the entrance to the walk, which circles around Tempe Town Lake, the sponsors have tables for their businesses. Many, but all, of these have a lot to do with the medical industry whose business helps diabetics in general. I toured the event with my father and wondered aloud: "Where is Chase?" I could not be the only employee in this company with the disease. Here sits a major event within a mile of one of their major employment centers and there is no presence to be felt. That had to change. Professionally I pursued this matter with the head of Chase's charitable giving department. After that conversation it did not take long for my message of hope to be spread among my own team members. My supervisors allowed me to take time during our monthly staff meeting and explain to these folks what it was that I was after and that their help would be greatly appreciated in future walks. Needless to say I have been back to four walks since. Chase now sponsors a table at each event. I don't know if that has anything to do with me and I don't care. The point is that in some small way I was able to make a difference.

On the morning of October 9, 2004 I received a call about 7:30 AM. I was already at work on this Sunday morning when the pager went off yet again. This time, I was in Maureen's office with great news. There were now two viable organs available for transplantation! If one would not fit me the other one will. This time I was prepared already. After the earlier near miss I went back to Wal-Mart and prepared a suitcase full of medications, clothing, and other such items so that when the phone rang all I had to do was grab the suitcase and go.

As our airplane pulled into Dulles International Airport it was between eight and nine o'clock in the evening. Upon landing I told my dad that this was for him as much as it was for me. I thanked him dearly for allowing me the strength necessary to fight this killer, stare it down in an alley, and conquer it. I was a very relieved man as I checked into the hospital that evening. I slept for approximately eight hours before the organ testing was completed. I was then wheeled into surgery. Before I left the room I gave my father Deb Harvey's business card with her phone number. The only reason it was hers is because I didn't have the business cards for two bosses which were mine at the time. I asked him to call her with the news when I went into recovery. This company had so much class it's not funny. When dad told them of the news they put it on the company ticker in Tempe. Peers and coworkers of mine were overcome with emotion as they tell me. They told me of having to take extra breaks that they may shed tears of joy over the fact that one of their own had beaten a killer. This spirit went so far as to have the sister of one of my coworkers show up in the hospital unbeknownst to me! To have her leave me a bouquet of flowers in the hospital was as if the employees themselves had left it. I will always be a supporter of transplantation for the purpose of saving lives. I continue my support of this cause.

Upon return to Phoenix I needed Christmas off. Rather than rush back to work I waited until January 3, 2005, to make my return. When I did come back the rules I had been accustomed to changed again. This now made the job even more difficult to do. I had to complete average calls faster than before. Along with this I realized I should have waited until at least May of that year to return to work. Any time one goes through a traumatic event like I did they must take time to recover and make sure that their mind is ready as well as their body for the return to regular work. I ran into trouble dealing with a heavy load of medicine designed to keep the pancreas from being rejected by the body. In addition, there were drugs to prevent virus, fungal infection, and the like. A side effect of all that medicine at once was an inability to stay awake for extended periods of time. That sounds like it doesn't make any sense when you think about the type of job I was holding. Nevertheless I found myself slipping further and further into a cycle of depression. This happened because my performance on the job was suffering. I can remember being told that I was not the man they thought I was. My reliability had been snared by someone who did not believe I had it. Once one loses managerial support he has to know he's on slippery ice as an employee. One of my two supervisors asked me once to quit. I frankly told her that she did not know me very well. I never quit! If you want to get rid of me fire me! I had taken this far enough to place me in front of Human Resources. They compared my pancreas transplant to having a baby! Please! You can plan a baby! Pancreatic transplants are not planned in advance. They may be set up in advance but then it becomes a waiting game. One doesn't know whether they're going to receive the transplant tomorrow, three months from now, three years from now or ever. To compare that to giving birth is absurd!

About four weeks later I was getting into a heated conversation with a customer who was irate over her credit card issue. Despite repeated attempts to soothe her emotions with a commonsense explanation of what she could do to solve the problem I got nowhere. I hung up at the end of the call, broke down in tears and asked for my release from the job. Instead I was suspended without pay. Four days later I was called in and unceremoniously fired by the company. I asked the appropriate questions I needed to regarding any kind of a reference, retirement plan, and the like. Having been told that it took five years seniority to be vested in the company, I was out there too. Had I managed to survive seven more months I would have been vested in Chase's stock program. This would have been my career. This would have been my retirement. Instead when they asked me for final comments I told them that in my book we were even. They had exposed me to a side of corporate business that I had never experienced before. In return they rewarded me by giving me a chance to transplant an organ in my body which would solve a riddle that had haunted me for over thirty years. When they offered me the opportunity to go to their employment bank for openings I told them no thank you. I thought that I would have a job so fast it would make your head spin. I was fired on a Tuesday. Saturday I had a resume turned in with the State of Arizona Department of Economic Security. Within a week after that application I passed their entrance test. I followed it up with an interview and passed with flying colors. The background check took an agonizing two months. During this time I exhausted much of the savings I had built up through Chase. But the job was mine by August. It's pretty amazing what a man can do when in a crisis. This trick was turned in four short days.

In closing, should my former employers ever get a copy of this book, I want to thank them all for their willingness to work with me in spite of challenges. Though there were times we had disappointments, I do not harbor them. Instead I choose to look back on the positive things that came from them. Without your display of courage none of this happens.

The Current Situation: 2005 – Present

In August 2005, I returned to work with the state of Arizona as an Eligibility Worker with the Arizona Department of Economic Security. To those of you who have been reading it sounds like a rerun of a job I held in 1991. It is the same job. This time I came with a better sense of organization and a willingness to stay ahead of the game as compared to falling behind while trying to decide 60 clients' fate in two days' time. The training required before one took this job on a daily basis was by and large the same as before. There were slight differences with respect to drug laws, possession limits, and the like. One thing that was not gone was the desire to help others help themselves.

Once I arrived on site I had the luxury of being able to perform an interview on the telephone in a client's personal absence. In essence this allowed me to inform a client exactly what they needed without having to be in my office. On the positive side of the coin their fuel expense was eliminated. On the negative I no longer saw a client in person during an interview. This makes identification extra difficult. Fraudulent identification cards can be copied and sent through the mail just as easily as real ones can show up on your desk. I can remember one incident where I had a client with a personal interview. This gentleman would have qualified for food stamps under any other condition except the fact that he had previously served prison time for drug possession and sale. By recent, I mean this gentleman had just been released from jail within the past three months! I don't know if any of you have ever felt the fear of telling an ex-convict "no", but when I sat in an electric wheelchair plugged into a wall and had to say that word, for just that split second I feared for my life. My heart went out to the man because he was trying to make a brand-new start in a world which would otherwise reject him. Other clients had visited me and left expired visas as identification. Had my eyeballs noticed it at the time this would not have been a problem. My failure to do so enabled the client to receive her food stamp benefits for an extra two months while we went through a hearing process. This is an example of just how exacting government work has to be.

Under these new rules I was able to maintain a pretty good workflow for approximately six months. In the middle of those six months illness struck again. As you may know by now I have been a Phoenix Valley resident for thirty-five years. Suddenly I developed flu-like symptoms. Approximately three days later I returned to work. Fortunately the symptoms occurred on a Friday. This allowed me to take a weekend and recover. Within 48 hours I began to develop symptoms of pneumonia. I had also developed an open sore on the job. The combination of these two factors landed me in the hospital where I spent the next 25 to 30 days. During this timeframe I lost 30 pounds. I was tested for every illness under the sun including lymphoma, a cancer of your lymph system, which is designed to fight infections. I was petrified. One night in the hospital bed I lay waiting for a spinal tap. With all the challenges I have already explained the last thing I needed was cancer! The spinal tap was not done. Instead another routine lab revealed Valley Fever as the culprit. I was shocked! I thought those kinds of illnesses occurred with newcomers to the Phoenix area. I guess I should've known that with a kidney and pancreas in this body that were not mine, any infection could form its own beast.

Being this ill I stayed on top of my situation by contacting my employer frequently. The last thing I wanted to do was lose a new job as a probationary employee. Upon return in late February I took back my workload from the workers who so nicely filled in for me in my absence. Once I cleaned up what they had done I took off on my own caseload. About a month or so into that the faith my management had in me began to erode.   It was not over the quality of the work that I had performed. Rather, they just wanted more of it. They needed workers who could average six to seven interviews per day while I could only give them four. I can remember my boss telling me that they appreciated the work I was doing but the pace had to be picked up or the job would eat me alive.

The joy of returning to work finally ended in May, 2006. While home with my father over a weekend I again became ill enough that dialysis was required to keep me alive. Reluctantly, I handed in my resignation. I can't remember how quickly I went to dialysis but what was done was done. I spent approximately seven months in the Fresenius medical care unit in Chandler, Arizona.

The predominant feature of this clinic that is different from anything I've ever been in was the extreme cold. I can remember being in other clinics both as a patient and a worker where it felt like Miami Beach compared to this place. Many a night it took three blankets to keep me warm throughout the treatment. I would often preface the treatment with two hot and spicy chicken burritos just to warm my insides! If you don't believe me, ask some of my current staff who have been in that Chandler clinic about just how cold it is.

February 18, 2007 will always be a memorable date for me. This is when I transferred to a brand-new Fresenius facility in Queen Creek, Arizona. I could not have been happier! Now I can get my dialysis within five miles of home. A seven minute drive will save a lot of wear and tear on vehicles and my body. This clinic is Miami Beach compared to the one in Chandler. This does not mean that it is not cold. I prefer 65° over 55° any day of the week. Even at this increased temperature clients like me have complained to staff over the years about just how cold it was. They now set their thermostat temperature at 72°.

Temperature is an important factor in dialysis treatment. Earlier I explained that keeping it cold is how germs are kept to a minimum and patients do not get sick as easily as they would otherwise. I doubt that the temperature will rise again. I had many challenges once I arrived in this new clinic. I realize now that most of them were not necessary and self- inflicted.

My laboratory numbers that February were off the charts in the wrong way. Being so young I made a vow that it would be turned around in all areas. To many staff at the time I was in denial, crazy, or living in some kind of a fantasy world. Often times these views were met with emotional outbursts by me. I had other patients telling me that they did not appreciate it. One, who shall remain nameless, went as far as to tell me: "Give it a rest, man! You have already had your chance!" Boy was this gentleman sadly mistaken.

He is still alive at another clinic within our system. Sir I only wish you the very best of luck in your battle with kidney failure. But if you ever read this book I only wanted you to understand that in this life there is only one chance. I will not cut my chance short because God gave me a couple of gifts along the way. Instead I will live every day of my life like it is the last day for one day it will be.

This type of behavior continued when I tried to deal with staff. I tried to deal with dietitians who give me the advice: "You need to eat more." There was other such advice along those lines over the next 3 to 4 1/2 years. Along with the foot check controversy that I brought up earlier it was enough to make a man cry. Often it did. Couple this with my other aforementioned weakness. Constant turnover within dialysis staff created a revolving door style of management and technicians. I have learned that change does not matter if the service the professional provides is the same. To all of the other professionals that I have consulted, cried about, or was very angry with for no apparent reason other than personal pride and stability, my sincerest apologies. If you've gotten this far into the book electing to get rid of it over this memory, you don't know what you're missing!

After several steps of contrition, and a severe reality check, staff, and patient had turned around severely. Along the way there were two more hospitalizations to get to this day. The first happened in the late winter of 2007. For reasons unbeknownst to me I could not bring my blood pressure down throughout the day. Even after dialysis I was running in the 150s over 80s. My father was at work at the time and had no idea just how serious this might get. I did not either. Within four hours I could not complete sentences talking to my own father. I knew it was time to go to the hospital without my father even suggesting it.

I spent approximately the next week at Scottsdale Healthcare Osborn in a variety of hospital rooms and settings. As it turns out my blood pressure was high enough that they had to put me on liquid Clonidine. Like many other drugs, when administered through an IV line, they will hit your bloodstream directly. This caused hallucinations, an extreme sense of loss of control, and in general, a very lousy experience until they were taken away. There are at least two different occasions where I thought death was at hand. I was envisioning a place I was taken while tests were run where I would grab my own shotgun and put an end to the whole thing. Another such hallucination was that of being involved in my own funeral from the casket! The situation was not made any easier when my former nephrologist, through all my years as a diabetic, literally stood in my hospital room with his partner that night and made reference to the fact that he thought I would be dead by morning and not to worry about me! One week later I was out of the hospital. Free from the drug in its vicious form I gradually began to come out of this. When I tell you gradually, I mean I could not tell what day of the week or what time of day it was for roughly a week following release! Once this cloud cleared I soon followed up with my nephrologist. Dad was with me at the time but that did not soften my message. I told this doctor as I have many others, *"DO NOT WRITE ME OFF AGAIN."* Having been my nephrologist for that long he should've known me better than he did.

A second incident occurred February 23, 2008. On my 41st birthday, a sleepy 16-year-old female driver nearly claimed my life, along with that of every other occupant of our vehicle. I had just been down to Tucson following an earlier dialysis run within our clinic. During the wedding reception I was asked to dance with a particular lady. I turned her down because I felt weak after a day of dialysis and travel. Because of turning her down, we were involved in a near head-on collision. That's the way my brother would say it. Actually he may be right. As I remember it I was ejected backwards from the passenger seat and landed on my feet like a gymnast in the rear of the SUV. The second I landed I heard a pop in my right hip. Immediately I knew that leg was broken and I begged the second one didn't go with it. From then on my body went into shock at such a level that I couldn't remember anything until I was dragged into a gurney which was hooked to a waiting helicopter. I do remember that my family and I went to separate hospitals. The insured folks went to University Medical Center in Tucson. I chose to go to Scottsdale Healthcare Osborn for the simple reasons that they had my kidney transplant records available on file and that my nephrologist was right across the street at his practice.

Despite the fact that we were all injured to varying degrees no one was killed. I consider that a miracle in itself. I took the worst of the injuries for the simple reason that I did not have a seat belt on at the time of the accident. This would normally have been common practice for me. I tried to put one on as we left the wedding reception but declined it because of the difficulty fitting the belt around my swollen body parts.

The next month was a kaleidoscope of activity, surgery, dialysis, rehabilitation, and hospital food which I really did not care about. I think I was under the control of morphine and any other painkiller staff could bring me when I needed it. In my lucid moments I progressed from a liquid to a solid diet. I even attempted to walk on many occasions. So the time came to release me to a rehabilitation unit which was completely apart and separate from the hospital. This unit was much closer to Mesa than to my own home. In fact I thought I was a jinx at first. As it turns out within 48 hours of my arrival my initial roommate passed away! At this point I wanted out of there so badly that I could taste it. For another two to three weeks, (time is not memorable), I underwent dialysis and rehabilitative therapy, just as I had in Scottsdale. The primary focus of this unit was rehabilitation after the accident.

When I returned home I had outpatient therapy in front of me. For the next 15 months I spent two days every week in the offices of a rehabilitation therapist group. During this time I began to walk between parallel bars, around parallel bars, up and down the carpet away from the parallel bars, and perform exercises on pads with help from a therapist. Just when I thought I would become depressed over the process a nine-year-old girl would show up with a sprained ankle or an ASU ballplayer would show up with a ruptured tendon in his leg. Despite their injuries I did not see any of them cry. I knew right then I had to cheer up and go through it. Gradually the continuing toll of stepping forward on my right leg became more severe. Many treatments would come where I had to take two painkillers within three hours of each other. One of them was taken about one hour before treatment and the second was taken during treatment. One such instance knocked me to the floor during therapy. When your kidneys fail medicines take longer to take effect. Because of this they also last in one's system longer than most. Now I hardly ever take a painkiller unless it is absolutely necessary. After these 15 months the business made a decision that I would no longer be helped by their services. This was not because I did anything wrong but because in their view they could not progress with me any further. So there I sit confined to a wheelchair. It is now August of 2009 and I have no hope to lean on. In the next year I found myself in the hospital twice more. Both instances related to pneumonia. By January 2011 I was recovering from glaucoma and purchasing glasses to wear for the rest of my life. Once all of these items are added to the rest of my health history I start to hate middle-age.

I decided to make a turnaround of another type. Shortly after leaving the rehabilitative hospital in 2008 I turned to the State of Arizona Division of Developmental Disabilities. This agency is the enforcement arm of the Social Security Administration's disability payment system. They are state-based which allows them to help the federal government release their funds as part of a joint agreement between the State of Arizona and the federal government. The funding mechanisms used to provide this service are split between the two entities. For me to explain this here would be way too much information in such a short book.

You could say that I was going stir crazy. Being trapped in my own home for no other reason than an inability to walk and dialysis treatments three times a week just didn't seem right for a man with my background and ability. In May of 2010 this agency hooked me up with a local organization specializing in day treatment of physically, emotionally, and mentally handicapped adults. Once I was contacted by Patty, Chuck, and the rest of their ownership group, a volunteer placement was arranged with Independent Life Services, Incorporated. This arrangement allowed me to work as a volunteer for several hundred hours per year as part of a contract between me, the agency, and the state of Arizona. Being a volunteer has its pluses and minuses. Among its pluses is the ability to freelance one's day on the fly. As the agency schedule changes I must be able to adapt mine. Within the first six months I spent the majority of my time learning. Again this meant learning who was who, what their individual roles were within the organization, what each client's ability to speak and learn was, and how to develop a goal based upon observation alone. This skill is very important when you don't have the ability to read confidential information. It is of utmost importance to know an organization's management structure in any setting! If a volunteer has no authority, that volunteer had better know who the bosses are, and take his direction from them.

Over the first six months I was dipping my hand in a few different projects. Primary among those was resume writing for the higher functioning client. After trying this with roughly 1/3 of the then current client base I quickly realized my limitations. The key struggle that our clients have will vary from person to person. Most important, their mental focus and comprehension ability is far less than that of a normal human being. This does not mean that they cannot learn. Rather it means that they are in different stages of learning emotional control, physical muscle strength and development, cognitive ability and so on. This means that every day is a different day.

One day in 2011, a new client came to ILS with a mission all her own. This young lady in her early 20s has issues that I will never be able to detect. It hasn't stopped the two of us from working on her dream. I have so much support from staff with her that I feel I cannot lose! All we have managed to do in approximately eight months' time is develop a working document she may use to look for a job. To further challenge her ILS has converted its store within the program to the point where she is running it. She has the ability to count, collect, and return small sums of money for items purchased by the clients within the store. Our store conducts rummage sales of its items. They are largely donations from the general public of items that were once used by others. We then turn around and sell the items at deep discounts to community members in need. Other activities include car washes for donation only, agency breakfasts and lunches, and outings for the clients which give them an opportunity to go out into the community. All of these things generate revenue and require revenue in order to continue functioning properly. In these ways the clients are making their own revenue. In turn, the cost of their outings per individual is lower. Our community sees what we are doing and offers outings at a reduced price to our clients. Many times, our activity is simply that of going to a public park. This single afternoon allows clients to have a meal with their peers, be active on a playground and a variety of different games, and just laugh. Within the next 12 to 18 months at my estimation, I am hoping that this young lady has her job. Once she is able to make at least some money every paycheck it will be hers to spend in the way she wants to. It will be hers to save the way she wants to. If this is not a story of human success I don't know what is.

This lady is an example of only one element of the ability I can bring to a table. One of their favorite activities is music day. One of our staff members has quite a bit of musical equipment which he brings from his own home as a tool to encourage the clients to exercise. They enjoy the day so much that it is not considered exercise. This has blossomed to the point where other staff members will fill in for Mark when he takes professional development classes. With any luck at all I will be able to join forces with Mark for at least one show before my time is done. Anyone in the area who would like more information on the program should give them a call.

A second prong of volunteerism is allowing me to maintain my mother's promise beyond maintenance. I am currently involved with Donate Life Arizona. This organization serves to raise awareness regarding organ donation throughout the State of Arizona. Their main goal is that of saving lives! By urging Arizona residents to register as organ donors they help to alleviate the supply and demand problem that exists. As I write this book there are over 2000 people in Arizona awaiting the gift of life. I am one of these people. With our family history my father and I felt it was only appropriate that we join this organization. This was a tough one for me as well. For the first six months of my involvement I was overwhelmed with statistics. When the writer himself is part of that group waiting for a miracle how can he keep objectivity? Every idea I had seemed to be shot down early on. It was almost as if I felt I didn't belong at all.

Then in January of this year our director got a memo from his bosses asking him to come up with 500,000 new registered donors within the State of Arizona by the end of 2012! This is an awesome undertaking for one person let alone the over 60 that we have in our chapter. I knew that this was the time I could ask the same proposals that I had the previous summer if I only counted correctly. Using the contacts I have, and some that I have yet to make, I asked him to help me work with the University community and get their involvement in our cause. I figure that with father and I as a team we can provide a powerful one-two punch showing the importance and the impact of the gift of life. As of release time Arizona State University is still developing a program. If successful the program will aid Donate Life Arizona in its cause! If this is not a great lesson in patience, show me a better one. There will be more information in this book intended for Arizona readers on how to get involved if they wish.

One of the greatest lessons you can learn about employment and responsibility comes when you volunteer for an organization first. If one can show a supervisor action, passion, commitment and respect for their goals, the supervisor may very well reward them with a job when available. I have landed jobs on two different occasions using such a method. I earned the others the old-fashioned way. All this does is show an ability to use plan B when plan A doesn't work. I have had a lifetime of this kind of thinking. Finally it seems that plan A is working first. After a recent foot sprain I spent three weeks in either a cast or an ankle brace. Once the cast came off I immediately put into motion a plan to get myself on my feet again. At release of this book I will be involved in the middle of six weeks of rehabilitative therapy. When the professional that is working with me is finished I will be able to stand behind a walker! The freedom of being able to move without support from another human being is nothing short of staggering. I only hope to be able to do this as baseball season opens.

<u>Epilogue</u>

In closing, the message of this book is simple. Throughout life all of us have tests of faith. It is up to each of us individually to respond to our challenges in the way that we see fit. The advice given in this book is strictly that. Consult any and all professionals where necessary. All solutions mentioned in this book are merely an option and should be considered as such on a case-by-case basis by individuals. The book is meant as a self-help tool. To my family and friends consider this my gift. We have all gone through crises of our own. It is our choice as to how we cope with them. In your consideration please take pieces from me as an example of how to overcome your obstacles.

THERE IS A VIABLE SOLUTION FOR TYPE I DIABETES

The following happened 11/9/2004 at the University of Maryland Medical Center. My life was permanently altered on that day for the better. It followed my 30-year path of hereditary Juvenile Diabetes. This is the same Type I diabetes that resulted in the death of my mother after 50 years of fighting the disease. Based on my experience in 2004 and since, I can recommend a Pancreatic Transplant.

How do you attempt to obtain a pancreas transplant?
1. Discuss the option with your Nephrologist, or health team.
2. Check with your employer's Benefit Coordinator to determine coverage level.
3. Be age 21 or over and prepare to meet the transplant program criteria.
4. Surgeries are performed in a number of major cities throughout the United States. Use your Benefit guide to find the program most beneficial to your survival/recovery.
5. Make your selection and attend a Transplant Education and Workup day at their hospital campus.

While you are on the waiting list you must leave the Transplant Coordinator a 24/7/365 contact number. In the interim continue to send monthly required labs as directed by your transplant team.

Once you have this wonderful gift take care of it by never forgetting anti-rejection or other prescribed medication.

In closing my experience says to check with your benefits coordinator where you work to see if this is a viable option for you.

If you live in the State of Arizona and would like to become a registered Organ or Tissue Donor contact 1-800-94-DONOR, or go to www.donatelifeaz.org. If you live out of state you may go through your state's network.

So You Want to Win the Climb to 4.0
(Normal Albumin Level)

The following advice is for anemic patients who want a quick increase in Albumin and Hemoglobin levels. Hemoglobin levels around 11.0 will fluctuate monthly. If they fall below set levels your dialysis nurse will inject EPO into your line during treatment. This level changes less rapidly than others. If it falls too low you may be sent for a blood transfusion.

As all of us know protein deficiency is one of the key problems of kidney disease. How many times in your treatment have you heard the dietician tell you that you needed to eat more protein? I heard it for over five years. The moral of this handout is that we can all get to (4.0) protein levels and renal meals do not have to be boring! As a bonus, you will see tips which will help bring other lab values into the normal range. The approach used in this handout pulled my albumin level from 3.5 as of August 15 to an astounding 4.1 by October 17. (See my lab results chart at the bottom.)

Typical Daily Menu:
To the right of each meal/item is listed the approximate number of protein grams.

Breakfast: 31 grams of protein
One "Zone" Protein Bar of your choice.   (15 g)
One "LiquaCel" protein supplement mixed in a cup of hot tea.   (16 g)

<u>Lunch:</u> **12 to 25 grams of protein**

> The idea for beginners here is variety. Any high protein meat sandwich will help.

<u>Dinner:</u> **25 grams of protein**

> Have any of the four major meats and a vegetable. (3.5 ounces of meat = 25 g)

<u>Dialysis:</u> **15 grams of protein**

> Because you lose protein during dialysis bring a high-protein snack to eat during the last half of treatment.

This handout is also intended to help patients defeat other problem chemicals. To handle high Adjusted Calcium (9.8+), use low (10% calcium) Zone protein bars like the Cinnamon Roll flavor instead of higher (20% to 50% calcium) protein bars or drinks. Better still, ask your Charge Nurse about adding LiquaCel protein supplement to a hot cup of tea daily. (16 grams)

If phosphorous levels are high (5.6+), try non-carbonated beverages like iced tea, fruit punch and lemonade flavor drinks. If you prefer carbonated drinks stick with root beer or a choice of 7-up or Sierra Mist. Take Phosphorous binders as directed by your physician and watch those numbers fall!   As they do your need for as many binders should decrease as well.

If your concern is potassium levels above 6.0, try non-citrus juices like grape, apple or cranberry. Avoid potatoes! Vegetables to avoid include tomatoes, broccoli, and spinach. Fruits to drop include all citrus, melons, and bananas. Coffee lovers are in for a 700 mg. infusion of potassium for each cup of coffee. Help alleviate this with a hot tea or any of the above suggested drinks.

This section is for Diabetic Patients. Please follow your Physician's advice. A1C levels higher than 7.0 are dangerous. Having been a Type I diabetic I can tell you that for every 0.1 rise above 7.0 there is an increased risk of developing heart disease. This can be controlled with better blood sugar control.

Your improvements will vary. If you have any questions, ask your dialysis staff and dietitian for details.

Author's Dialysis Lab Results:					
	Aug. 2011	Oct. 2011	Dec. 2011	Jan. 2012	Goal
Albumin	3.5	4.1	4.5	4.7	4.0 +
Hemoglobin			11.4	11.6	10.0 +
Adjusted Calcium			10.0	8.8	8.4 to 9.5
Phosphorous			4.6	5.1	3.0 to 5.5
Potassium			4.6	3.6	3.5 to 6.0
A1C			4.6		under 7.0
Kt/V		1.36*		1.99	1.20 +
* (I increased to the 180 Filter at 3.5 hrs. run time.)					

Your fellow Patient,
James Schaner

www.ingramcontent.com/pod-product-compliance
Lightning Source LLC
Chambersburg PA
CBHW022115170526
45157CB00004B/1653